T0177835

The 9 Pitfalls of Data Science

Gary Smith and Jay Cordes have a most captivating way and special talent to describe how easy it is to be fooled by the promises of spurious data and by the hype of data science.

Professor John P.A. Ioannidis,
Stanford University

Smith and Cordes have produced a remarkably lucid, example-driven text that anybody working near data would do well to read. Though the book is presented as fables and pitfalls, a cogent, scientific approach reveals itself. Managers of data science teams stand to learn a great deal; seasoned data scientists will nod their heads knowingly.

D. Alex Hughes
Adjunct Assistant Professor, UC Berkeley School of Information

The current AI hype can be disorienting, but this refreshing book informs to realign expectations, and provides entertaining and relevant narrative examples that illustrate what can go wrong when you ignore the pitfalls of data science. Responsible data scientists should take heed of Smith and Cordes' guidance, especially when considering using AI in healthcare where transparency about safety, efficacy, and equity is life-saving.

Michael Abramoff, MD, PhD,
Founder and CEO of IDx
Watzke Professor of Ophthalmology and Visual Sciences at the
University of Iowa

In this era of big data, it's good to have a book that collects ways that big data can lie and mislead. This book provides practical advice for users of big data in a way that's easy to digest and appreciate, and will help guide them so that they can avoid its pitfalls.

Joseph Halpern
Joseph C. Ford Professor of Engineering,
Computer Science Department, Cornell University

Increasingly, the world is immersed in data! Gary Smith and Jay Cordes offer up a veritable firehose of fabulous examples of the uses/misuses of all that "big data" in real life. You will be a more informed citizen and better-armed consumer by reading their book…and, it couldn't come at a better time!

Shecky Riemann
math blogger

THE 9 PITFALLS OF DATA SCIENCE

GARY SMITH AND JAY CORDES

Great Clarendon Street, Oxford, OX2 6DP,
United Kingdom

Oxford University Press is a department of the University of Oxford.
It furthers the University's objective of excellence in research, scholarship,
and education by publishing worldwide. Oxford is a registered trade mark of
Oxford University Press in the UK and in certain other countries

Published in the United States of America by Oxford University Press
198 Madison Avenue, New York, NY 10016, United States of America

British Library Cataloguing in Publication Data

Data available

Library of Congress Control Number: 2019934000

ISBN 978-0-19-884439-6

DOI: 10.1093/oso/9780198844396.001.0001

Printed and bound by
CPI Group (UK) Ltd, Croydon, CR0 4YY

CONTENTS

Introduction		1
1. Using Bad Data		3
2. Putting Data Before Theory		33
3. Worshiping Math		65
4. Worshiping Computers		85
5. Torturing Data		111
6. Fooling Yourself		127
7. Confusing Correlation with Causation		155
8. Being Surprised by Regression Toward the Mean		173
9. Doing Harm		197
Case Study: The Great Recession		229
Bibliography		241
Index		253

Introduction

A 2012 article in the *Harvard Business Review* named *data scientist* the "sexiest job of the 21st century." Governments and businesses are scrambling to hire data scientists, and workers are clamoring to become data scientists, or at least label themselves as such.

Many colleges and universities now offer data science degrees, but their curricula differ wildly. Many businesses have data science divisions, but few restrictions on what they do. Many people say they are data scientists, but may have simply taken some online programming courses and don't know what they don't know. The result is that the analyses produced by data scientists are sometimes spectacular and, other times, disastrous. In a rush to learn the technical skills, the crucial principles of data science are often neglected.

Too many would-be data scientists have the misguided belief that we don't need theories, common sense, or wisdom. An all-too-common thought is, "We shouldn't waste time thinking about *why* something may or may not be true. It's enough to let computers find a pattern and assume that the pattern will persist and make useful predictions." This ill-founded belief underlies misguided projects that have attempted to use Facebook status updates to price auto insurance, Google search queries to predict flu outbreaks, and Twitter tweets to predict stock prices.

Data science is surely revolutionizing our lives, allowing decisions to be based on data rather than lazy thinking, whims, hunches, and prejudices. Unfortunately, data scientists themselves can be plagued by lazy

The 9 Pitfalls of Data Science. Gary Smith and Jay Cordes. Oxford University Press (2019).
© Gary Smith and Jay Cordes 2019. DOI: 10.1093/oso/9780198844396.001.0001

thinking, whims, hunches, and prejudices—and end up fooling themselves and others.

One of Jay's managers understood the difference between good data science and garbage. He categorized people who knew what they were doing as *kids*; those who spouted nonsense were *clowns*. One of our goals is to explain the difference between a data scientist and a data clown.

Our criticism of data clowns should not be misinterpreted as a disdain for science or scientists. *The Economist* once used a jigsaw puzzle and a house of cards as metaphors:

In any complex scientific picture of the world there will be gaps, misperceptions and mistakes. Whether your impression is dominated by the whole or the holes will depend on your attitude to the project at hand. You might say that some see a jigsaw where others see a house of cards. Jigsaw types have in mind an overall picture and are open to bits being taken out, moved around or abandoned should they not fit. Those who see houses of cards think that if any piece is removed, the whole lot falls down.

We are firmly in the jigsaw puzzle camp. When done right, there's no question that science works and enriches our lives immensely.

We are not going to bombard you with equations or bore you with technical tips. We want to propose enduring principles. We offer these principles as nine pitfalls to be avoided. We hope that these principles will not only help data scientists be more effective, but also help everyone distinguish between good data science and rubbish. Our book is loaded with grand successes and epic failures. It highlights winning approaches and warns of common pitfalls. We are confident that after reading it, you will recognize good data science when you see it, know how to avoid being duped by data, and make better, more informed decisions. Whether you want to be an effective creator, interpreter, user, or consumer of data, it is important to know the nine pitfalls of data science.

Using Bad Data

"Where's Alex [the often missing Director of Analytics]?"
"That's the second most common question we get."
"What's the first?"
"Why is revenue down?"

Jerry showed up for work one morning and found chaos. Revenue had dropped 10 percent overnight, and no one knew why.

The analytics department scrambled for reasons and couldn't find any. There hadn't been any power outages, server crashes, or known bugs in their software, but revenue had plummeted in every country, on every web page, and for every ad. A dozen data analysts met in a conference room to study the numbers and find an answer.

After a long and frustrating discussion, Jerry looked at the minute-by-minute record of clicks on millions of web pages and noticed that revenue stopped everywhere at 2 a.m. and didn't resume until 3 a.m. What are the chances everything would stop at exactly the same time for precisely one hour? It's embarrassing how long it took a dozen smart people to figure out what happened: daylight saving time.

It's even more embarrassing that the very same phony crisis happened the following year. They didn't learn from their mistake, but there is a bigger lesson here. Useful data analysis requires good data. These data suggested a nonexistent problem because the data were flawed. Good data

The 9 Pitfalls of Data Science. Gary Smith and Jay Cordes. Oxford University Press (2019).
© Gary Smith and Jay Cordes 2019. DOI: 10.1093/oso/9780198844396.001.0001

scientists think seriously about how their data might be distorted by daylight saving time, holidays, leap years, or other factors.

Earthquakes on the rise

Good data scientists also consider the reliability of their data. How were the data collected? Are they complete? Why might they be misleading? For example, the United States Geological Survey (USGS) earthquake data in Figure 1.1 show an alarming increase in the number of magnitude 7.0+ earthquakes worldwide over the past century. Is the earth becoming more fragile? Either we have discovered an apocalyptic story of the earth breaking apart, or something is wrong with these data.

Don't worry. These are the earthquakes that were recorded each year, not the number that occurred. There is now a far more extensive worldwide network of seismometers than in the past, so many earthquakes that might have gone unnoticed decades ago now get monitored and logged.

These data cannot tell us whether there has been an increase in the number of earthquakes, only that there has been an increase in the number of earthquakes that are recorded.

If the data tell you something crazy, there's a good chance you would be crazy to believe the data.

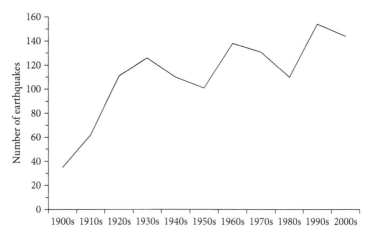

Figure 1.1 Magnitude 7.0+ earthquakes on the rise?

Stay indoors on the first of the month

The Chicago Police Department has made its crime statistics publicly available for data scientists to analyze. Suppose we want to identify the days of the month that had the most crime in 2018. Figure 1.2 shows that the first day of the month is crime day and the last few days of the month are crime vacation days! That sounds crazy, so a good data scientist would think about what might be wrong with these data.

One problem is that not every month has 31 days, so we should expect fewer crimes on 31st days than on 30th days, simply because there are fewer 31st days. Still, that doesn't explain why the number of crimes spikes on the 1st.

We couldn't think of a compelling explanation, so we probed deeper, separating the data into 18 categories representing the 18 most common types of crime. In 15 of the 18 categories, there were more crimes recorded on the first day of the month than on the second. Then we realized that these data are like the earthquake data. They are not a count of when crimes happen, but, rather, when crimes are recorded.

The date stamp on a crime record is not the day the crime occurred, or even the day the crime was reported to the police. It is the day when the

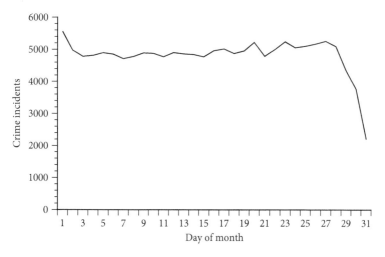

Figure 1.2 Most crime occurs on the first day of the month?

incident was entered into the police database. The data-entry people are evidently more likely to enter crime reports on the first day of the month than on the last few days of the month. We don't know why this is, but we do know that these data are useless for telling us whether criminal activity ebbs and flows over the course of a month.

Check your data twice, analyze once

A data analytics firm was hired to see whether people who served in the U.S. military earned higher incomes after they returned to civilian life than did people with the same education and standardized test scores who had not served in the military. The firm used a nationwide database and was able to control for age, gender, race, and other confounding factors. They were surprised when they did not find consistent differences in the annual incomes of veterans and nonveterans.

Then one of the team members looked at the data more carefully. She sorted the millions of observations into eight income categories and discovered an odd distribution like the one shown in Figure 1.3. A remarkable 37 percent had incomes below $25,000. Something didn't seem right, so

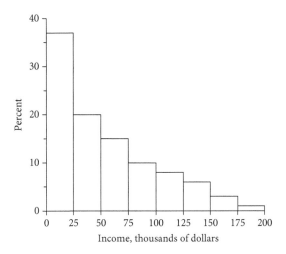

Figure 1.3 Income distribution, 2015

she sorted the incomes below $25,000 from lowest to highest and found that roughly half of the values were zeroes! Many people had either not answered the question about their income or were not working because they were homemakers, students, retired, or unemployed.

The team redid the analysis, this time looking at two different issues. Were veterans more likely to be employed and, of those who were working, were veterans paid more? This time, the answers to both questions were a resounding yes.

Sometimes, a statistical analysis falters because of clerical errors in the data, like a misplaced decimal point or an inadvertent negative sign. The Joint Economic Committee of Congress once reported that the fraction of household wealth held by the richest one-half of one percent of U.S. families had jumped from 25 percent to 35 percent. Some skeptics rechecked the calculations and found that the reported increase was due almost entirely to the incorrect recording of one family's $2 million wealth as $200 million, an error that raised the average wealth of the handful of rich people who had been surveyed by nearly 50 percent. It would have been better to catch the error before issuing a press release.

This $200 million outlier was a mistake. Other times, outliers are real, and data scientists have to think carefully about whether an outlier is an anomaly that should be discarded or is important information. For example, U.S. stock prices usually go up or down a few percent every day. On October 19, 1987, the Dow Jones Industrial Average dropped 23 percent. This was an outlier, but it was an important one.

First, it demonstrated that there was something terribly wrong with the mathematical models used throughout Wall Street that assumed that a Dow drop of more than 10 percent was practically impossible. A second lesson learned was that the Federal Reserve is willing and able to do whatever is needed to prevent a bad day on Wall Street from turning into an economic meltdown. The Federal Reserve reduced interest rates, promised to supply much needed cash, and pressured banks to lend money—saving the market on October 19 and during the days, weeks, and months that followed.

Outliers are sometimes clerical errors, measurement errors, or flukes that, if not corrected or omitted, will distort the data. At other times, they are the most important observations. Either way, good data scientists look at their data before analyzing them.

Hospital readmissions

Sepsis occurs when the human body overreacts to an infection and sends chemicals into the bloodstream that can cause tissue damage, organ failure, and even death. Jay worked with a group that used data for 7,119 patients with sepsis at a Chicago hospital to predict the chances of being readmitted to the hospital in the first 30 days after being discharged. Figure 1.4 compares the pH level of the patient's blood (normally between 7.35 to 7.45) to the hospital readmission rates for the seven pH groups for which there were at least 10 observations.

There is a clear positive relationship, indicating that patients with high pH levels are more likely to return to the hospital soon after being discharged. A low pH signals that a discharged patient is unlikely to be readmitted. The correlation is 0.96 and data clowns would call it a day.

Jay and his teammates were not clowns, so they asked a doctor to review their findings. When he saw Figure 1.4, a puzzled look came across his face: "That's strange; the relationship is backwards. If you have a low pH level, you're probably dead," but the chart implied that having a very low pH level was a sign of health.

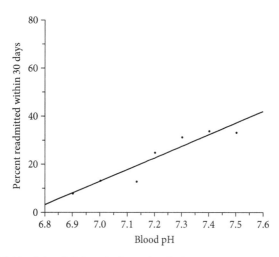

Figure 1.4 Hospital readmission rates for sepsis patients

This stumped the group until they realized that their data included patients who had died during their hospital stay! The patients least likely to be readmitted are the ones who were discharged to the mortuary.

Figure 1.5 shows that, once the team removed the deceased patients, the pattern reversed. Now there is a negative relationship, just as the doctor expected.

Figure 1.6 confirms the clear danger of acidic blood by comparing pH level with the likelihood of death. Contrary to the initial indication that patients with pH values below 7.2 are in relatively good health, they are, in fact, in serious danger.

In this age of big data and sophisticated tools, it is tempting to think that expert knowledge is optional. It's not. Data clowns, who plow through data and harvest conclusions based on data, data, and nothing but the data, risk embarrassing themselves by making ludicrous recommendations.

The young die young

In March 2015, a professor of psychology and music reported that musicians in traditional musical genres (like blues, jazz, and country) live much longer than do musicians in relatively new genres (like metal, rap, and hip hop).

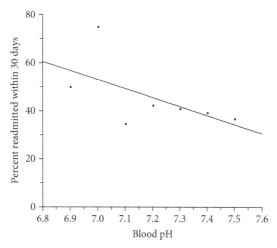

Figure 1.5 Hospital readmission rates for living sepsis patients

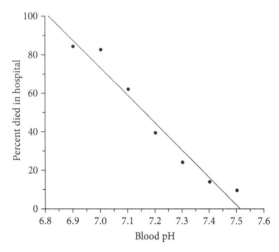

Figure 1.6 A low pH level really isn't good for you

Figure 1.7 is a rendition of the figure she used to document her conclusion, minus a piano player in the background. The professor concluded that performing these new genres was more dangerous than going to war:

People who go into rap music or hip hop or punk, they're in a much more occupational hazard profession compared to war. We don't lose half our army in a battle.

The average age at death for musicians in relatively new genres does appear to be dramatically lower than for musicians in traditional music genres and for the population as a whole. However, a closer look reveals some puzzling issues. The female and male life expectancy lines seem to be showing an increase in life expectancy over time, which makes sense, but years are not on the horizontal axis! The horizontal axis is 14 musical genres. Why are the trends in U.S. life expectancy graphed in comparison to musical genres, rather than by year?

It is also bewildering that the average age at death for rappers and male hip-hop artists is less than 30. Seriously, don't some survive until their 40s, 50s, or even older ages? If we pause and think about these data before comparing hip-hopping to going to war, the fundamental problem emerges. The new genres are so new that the performers haven't had a chance to live to an old age.

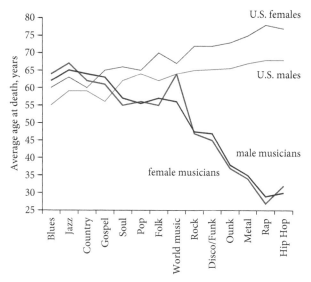

Figure 1.7 Age at death and musical genre

Hip hop began in the late 1970s. People who began doing hip hop in 1980 at age 20 and are still alive would be 57 years old in 2017. Anyone who died before 2017 would be younger than 57. People who began doing hip hop after 1980 would be even younger. In contrast, blues, jazz, and country have been around long enough for performers to grow old.

Once again, one of the keys to becoming a good data scientist is to use good data.

Let them eat cake

Figure 1.8 is an updated version of a *New York Times* graphic that accompanied an article by neoconservative David Frum titled "Welcome, Nouveaux Riches." The figure shows a dramatic acceleration between 1980 and 1990 in the number of households earning more than $100,000 a year. Frum wrote that, "Nothing like this immense crowd of wealthy people has been seen in the history of the planet." That sounds hyperbolic, but the figure does seem to show a huge increase.

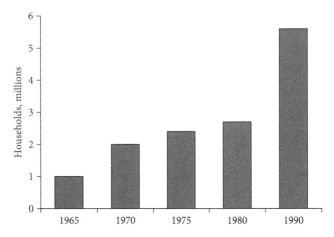

Figure 1.8 Number of households earning more than $100,000 per year

Do you notice anything odd about this figure? There is a 5-year differ-ence between each of the first four bars, but a *10-year difference* between the fourth and fifth bars (1980 and 1990). If the bars had been spaced properly and a 1985 bar inserted, the increase over time would appear gradual, without an abrupt jump between 1980 and 1990.

In addition, $100,000 in 1990 is not the same as $100,000 in 1965. Prices were about four times higher, so that $100,000 in 1990 was roughly equivalent to $25,000 in 1965. We should compare the number of families earning $25,000 in 1965 with the number of families earning $100,000 in 1990. We should also take into account the increase in the population. It is not sur-prising that more people have high incomes when there are more people.

Figure 1.9 fixes all these problems by showing the percentage of house-holds that earned more than $100,000 in inflation-adjusted dollars, with 1985 inserted, and data for 1995 through 2015 included to give more his-torical context. The 1980s are unremarkable. What does stand out is 2000, the height of the Internet bubble, followed by the burst bubble, the Great Recession, and recovery.

Good data scientists know that the purpose of a graph is to inform, not to mislead. Dollar data should be adjusted for inflation. Data that change as the population grows should be adjusted for population growth. Relevant data should not be omitted in order to deceive.

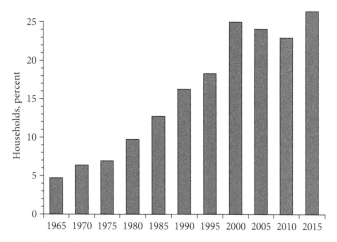

Figure 1.9 Percent of households earning more than $100,000 per year, adjusted for inflation

Law of small numbers

After a few bad rounds of golf, a weekend player uses a different putter and does better. It must have been the putter, right?

Maybe, but maybe it was just luck. Many outcomes are possible when an amateur tries to batter and coax a small ball into a little hole in the ground hundreds of yards away. As Winston Churchill once said, "Golf is a game whose aim is to hit a very small ball into an even smaller hole, with weapons singularly ill-designed for the purpose."

An amateur golfer deciding, based on a handful of putts, whether a putter makes a difference is an example of the error that Daniel Kahneman and Amos Tversky called "the law of small numbers." It is mistake to see something happen a few times (a few putts made, a few stock picks succeed, or a few predictions come true) and draw a general conclusion.

Even among professional golfers, luck is endemic. There is considerable happenstance in gusts of wind and in fortunate and unfortunate bounces. Sometimes a ball lands on a bank of grass and sticks; sometimes, it rolls into a lake or sand trap. Sometimes a ball whistles through a tree; sometimes it bounces off a branch. Sometimes a branch ricochet puts the ball back on the fairway, where the grass is cut short and the ball can be played cleanly; sometimes the ball bounces into foot-high grass. In a 2016 tournament, Phil Mickelson hit an errant drive and the ball caromed off the head of a spectator on one side of the fairway, and landed in the rough on the other side of the fairway. Mickelson told the spectator, "If your head was a touch softer, I'd be in the fairway."

A professional golfer who has the best score in the first round of a golf tournament is not necessarily going to win the tournament. A golfer who wins a tournament is not necessarily the best golfer in the world, and is certainly not assured of winning the next tournament he plays in.

In any competition—including academic tests, athletic events, or company management—where there is an element of luck that causes performances to be an imperfect measure of ability, there is an important difference between competitions among people with high ability and competitions among people of lesser ability. If four casual golfers play a round of golf, and one player is much better than the others, the winner is determined mostly by ability. If four of the top golfers in the world play a round of golf, the winner is determined mostly by luck. This is the *paradox of luck and skill*: the more skilled are the competitors, the more the outcome is determined by luck.

Let's look at some pro data. There are four major men's golf tournaments: the Masters Tournament, U.S. Open, British Open, and PGA Championship. In each of these tournaments, any of a dozen or more players might win. Who does win is largely a matter of luck. That's why there are so few repeat winners. As of February 2019, a total of 221 golfers have won at least one of the four major golf championships; 139 (63 percent) won once, and never again.

The Masters Tournament is the only major golf tournament that is played on the same course every year (the Augusta National Golf Club), which makes for a fair comparison of player scores in adjacent years. The Masters Tournament begins during the first full week in April with close to 100 golfers playing 18-hole rounds on Thursday and Friday. After these 36 holes, the top 50 players, plus any player who is within 10 strokes of the leader, make the cut, which allows them to play rounds 3 and 4 on Friday and Saturday.

There were 53 golfers who made the cut in 2017 and 53 in 2018, but only 27 golfers made the cut both years, which is, by itself, evidence of the importance of luck in golf. Figure 1.10 shows the 2017 and 2018 scores for the 27 golfers who made the cut both years. (Remember, low scores are better.) The correlation between the 2017 and 2018 scores is 0.14. There is a positive correlation, but it is loose.

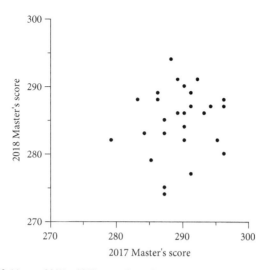

Figure 1.10 Masters 2017 vs 2018 scores for golfers who made the cut, correlation = 0.14

Figure 1.11 shows the 2017 and 2018 Masters scores for the top 15 golfers in these two years. The correlation is 0.03, essentially no correlation whatsoever. This is dramatic evidence of the paradox of luck and skill: the more skilled the competitors, the more the outcome is determined by luck.

Another way to see the importance of luck in golf is to compare scores from one round to the next. If performances depended entirely on ability, with luck playing no role, each golfer would get the same score round after round. Some golfers would get a 68 every round, while other golfers would get a 74 every time. If, on the other hand, performance were all luck, there would be no correlation between scores from one round to the next. A golfer with a 68 in the first round would be no more likely to get a 68 in the second round than would a golfer with a 74 in the first round.

To investigate which scenario is closer to the truth, we compared each player's third and fourth round scores in the 2018 tournament. Figure 1.12 shows that the correlation is 0.06, so close to zero as to be meaningless.

Good data scientists know that, because of inevitable ups and downs in the data for almost any interesting question, they shouldn't draw conclusions from small samples, where flukes might look like evidence. The next time you see something remarkable, remember how even the best golfers in the world are buffeted by luck, and don't fall for the law of small numbers.

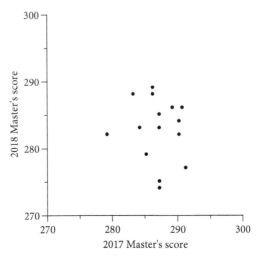

Figure 1.11 Masters 2017 vs 2018 scores for the top 15 golfers, correlation = 0.03

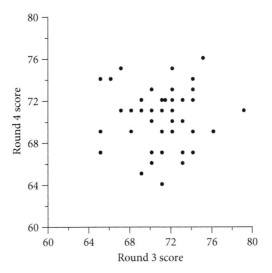

Figure 1.12 Round 3 vs round 4, correlation = 0.06

Small teams rule (and stink)

Let's do a little experiment involving the 537 players in the National Basketball Association (NBA) for the 2017–2018 season. We randomly assigned these 537 players to 100 "fantasy teams" by giving each player a random number from 1 to 100. There were an average of 5.37 players per team but, by the luck of the draw, some teams had more players than other teams.

We then compared the teams in 10 different performance categories. For example, Table 1.1 shows that Team 54 averaged 17.9 points per game, which was higher than any other team.

There were 5 players on Team 54. The average team size for the 10 highest scoring teams was 4.4, significantly smaller than the overall 5.37-player average. Table 1.2 shows that this was the rule, not the exception. In every single category, the average number of players on the top 10 teams was less than 5.37.

How could this be? Are there more personality conflicts on large teams? Is there more jealousy and in-fighting about playing time? Perhaps, but, remember, these are not real teams. They didn't actually play with each other; they were just assigned random team numbers.

One clue for making sense out of this puzzle is that Team 37 was the best team in two categories, averaging 90 percent shooting from the free-throw line and missing an average of only five games all season. The average player, indeed, the only player, on Team 37 is 39-year-old Dirk Nowitski. A team's average statistics are going to be pretty impressive if there is only one person on the team—and that person is a good player.

Let's look at the average size of the worst teams. Table 1.3 shows that the worst team in points per game had only two players and averaged a paltry

Table 1.1 *Team #54 averaged 17.9 points per game*	
Buddy Hield	13.5
D'Angelo Russell	15.5
Donovan Mitchell	20.5
Harrison Barnes	21.3
Blake Griffin	21.3
Team	17.9

Table 1.2 *Average number of players on top-10 teams*

	Average team size
Points per game	4.4
Rebounds per game	4.9
Assists per game	4.3
Blocks per game	4.1
Steals per game	3.9
2-point shooting percentage	4.9
3-point shooting percentage	5.3
Free-throw shooting percentage	3.8
Number of games played	3.8
Minutes per game	4.4

Table 1.3 *Team #59 averaged 0.8 points per game*

James Webb III	1.6
Ben Moore	0.0
Team	0.8

0.8 points per game. Overall, in 9 of the 10 categories, the 10 worst teams also had a below-average number of players.

That's the secret. The best teams are generally small, and so are the worst teams. There is nothing inherently good or bad about having a small team, but there is more variation in the average outcome for small samples than for large samples. If we flip one coin, there is a 50 percent chance of all heads and a 50 percent chance of all tails. If we flip two coins, there is a 25 percent chance of all heads and a 25 percent chance of all tails. If we flip 100 coins, there is almost no chance of all heads or all tails; indeed, there is a 96 percent chance that there will be between 40 percent and 60 percent heads.

It is not just coin flips. The average of a large sample is likely to be close to the overall average, while the average of a small sample has a much higher chance of being far from the overall average. That is exactly what happened with our NBA data. Because there is more variation in the performance of small teams, they dominated the lists of the best and worst teams.

This principle is subtle and often overlooked. Studies of school performance have found that the best schools tend to be small, suggesting that being small is a virtue. However, the worst-performing schools also tend to be small. Studies of cancer incidence have found that the lowest cancer rates are in small cities, suggesting that one can reduce the chances of getting cancer by moving to a small city. What's your guess about the size of the cities with the highest cancer rates?

Jay was recently congratulated by a friend for being one of the winningest players in the history of a well-regarded tournament backgammon club. Jay was puzzled, but then remembered that the friend had invited him to come to the club one time, and Jay had won three of four matches, which gave him a 75 percent win percentage, tied for 17th place and well ahead of the best player in California, who had a 63 percent win rate.

Seven players had perfect 100 percent win percentages. Four had played one match, one two matches, and two three matches. None of the top 25 players had played more than nine matches. At the other end of the list were 140 players with 0 percent win percentages. Only one had played more than eight matches. Again, those at the top and bottom of the list were the ones with small sample sizes. Comparing winning percentages, average test scores, or cancer rates without controlling for matches, school sizes, or population is a pitfall waiting to be stepped in.

Good data scientists are careful when they compare samples of different sizes. It is easier for small groups to be lucky. It's also easier for small groups to be unlucky.

A can't-miss opportunity

The April 4, 2016, issue of *Time* magazine contained a full-page advertisement touting gold as a can't-miss investment. Many of the claims were vague:

With predictions of the gold market rising past its record high price…individuals are currently moving up to 30% of their assets into gold.

One claim stood out as specific and verifiable:

If you would have taken $150,000 of your money and bought gold in 2001, then that initial purchase would have been worth over $1 million exactly 10 years later in 2011!

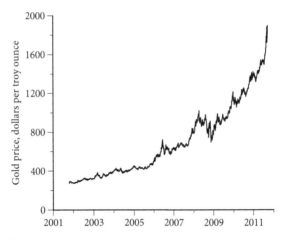

Figure 1.13 Up, up, and away

A footnote identified the period as September 6, 2001, to September 6, 2011, a period when gold did, indeed, soar, as shown in Figure 1.13.

Since the advertisement appeared in 2016, a vigilant data scientist should suspect that these gold enthusiasts chose this specific 10-year period in order to support their advertising claim that gold has been a great investment. Why would a study done in 2016 stop with 2011 data? That suspicion is confirmed by Figure 1.14, which includes several years before and after their cherry-picked 10-year period, stopping in April 4, 2016, when the advertisement appeared.

Good data scientists do not cherry pick data by excluding data that do not support their claims. One of the most bitter criticisms of statisticians is that, "Figures don't lie, but liars figure." An unscrupulous statistician can prove most anything by carefully choosing favorable data and ignoring conflicting evidence. Here, the gold hawkers made a clever selection of the years to begin and end a graph in order to show a trend that was absent in more complete data. In 2018 Nate Silver, the statistician behind the FiveThirtyEight website, tweeted that "cherry-picking evidence, at a certain level of severity, is morally equivalent to fabricating evidence."

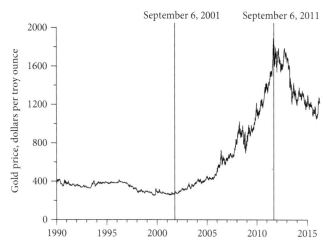

Figure 1.14 Gold always goes up, unless it goes down

Zero warming?

Our skeptic radar should be on high alert whenever someone uses data that begin or end at peculiar points in time. If an assertion is made in 2009 using data for October 2002 to January 2007, we should ask why they didn't use data for before 2002 or after 2007, and why their data begin in October and end in January. If the beginning and ending points are unexpected choices that were most likely made after scrutinizing the data, then they were probably chosen to distort the historical record. There may be a perfectly logical explanation, but we should insist on hearing an explanation.

In a 2015 interview on "Late Night with Seth Meyers," Senator Ted Cruz stated that, "Many of the alarmists on global warming, they've got a problem because the science doesn't back them up. In particular, satellite data demonstrate for the last 17 years, there's been zero warming." High alert! Why would Senator Cruz pick 1998, 17 years prior, as a reference point, unless he was cherry picking data?

Figure 1.15 shows annual global land-ocean temperatures back to 1880, as far back as these NASA data go. The data are shown as deviations from what the average temperature was for the 30-year period, 1951 through

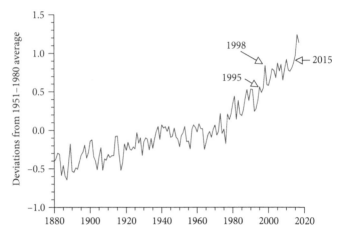

Figure 1.15 Global land-ocean temperatures, degrees Celsius

1980. It should be no surprise that there was a temperature spike in Senator Cruz's baseline year, 1998 (caused by a strong El Niño and unusually warm oceans). If Senator Cruz had used 1997, 1999, or the more natural 1995 (20 years prior to his 2015 interview), the data would not have supported his claim. That's why he selected 1998—to try to trick us.

Figure 1.15 also shows a rise in temperatures since the 2015 interview. There has been an increase in average global land-ocean temperatures over the past several decades, and it is egregiously misleading to cherry pick a 17-year period that misrepresents this trend.

If you can't convince them, confuse them

A Prudential Financial advertisement was titled, "Let's get ready for a longer retirement." This creative ad stated that,

A typical American city. 400 people. And a fascinating experiment. We asked everyday people to show us the age of the oldest person they know by placing a sticker on our chart. Living proof that we are living longer. Which means we need more money to live in retirement.

Figure 1.16 is a simplified version of their graph, omitting buildings, trees, grass, and other distracting, unnecessary decoration that tried to make

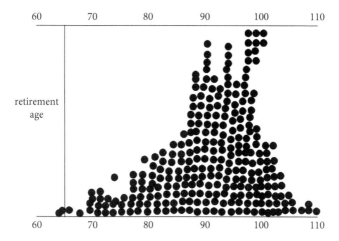

Figure 1.16 Are we living longer?

the figure seem like a frameable piece of art. The circles are a bit haphazard because they were placed on the board by ordinary people, adding to its folksy charm. The figure encourages us to think that these are the life expectancies of 400 people, most of whom will live past 90, so "we need more money to live in retirement."

Once again, let's think about the data before jumping to conclusions. These are the ages of the *oldest* persons these people know, not the ages at which typical people die. Perhaps many of the people who put stickers on the board know the same elderly people, especially in a small town. Using an extreme example to make the point, imagine that one person is 110 years old and everyone knows her. There would be 400 stickers at 110, even though none of these 400 people may live past 90. The fact that most people know somebody who is more than 90 years old doesn't mean that everyone will live to be more than 90 years old.

Where are the bad mutual funds?

A mutual fund salesperson showed Jay a list of mutual funds that he should consider investing in. The first thing Jay noticed was that practically all of them had beaten the S&P 500 index of stock prices. Unless they're from Lake Wobegon ("where all the women are strong, all the men are good looking, and all the children are above average"), that seemed pretty suspicious. Jay's skeptic radar went on high alert.

One bit of trickery is that poorly performing funds often disappear without a trace. An underachieving fund is absorbed by an overachieving fund, and the merged fund reports the past performance of the successful fund, as if the unsuccessful fund never existed. For a data scientist, this *survivorship bias* means that mutual fund performance data misrepresent the actual performance of the industry.

A valid study would consider all the funds that existed, say, 20 years ago and see how they performed subsequently—including funds that disappeared along the way. A retrospective study, in contrast, looks backward and considers only the 20-year performance of funds that still exist today. Retrospective studies are very popular in business, medicine, and many other fields because it is often much easier to obtain data for things that exist today than for things that no longer exist. The risk is that the results will be tainted by survivorship bias—the exclusion of businesses that failed and people who died.

For example, in 2018 it was reported that people who eat lots of cheese have fewer strokes and less risk of cardiovascular disease, but maybe many of those who habitually ate lots of cheese in the past are now deceased. The cheese-eaters who survived may have lower genetic risks, exercise more, or have other characteristics that offset their cheese eating. It is also possible that some people who do not eat cheese today were once big cheese eaters, but gave it up because of health problems. If so, they are more at risk, not because they avoid cheese, but because they once feasted on it.

One of the great stories of survivorship bias—one with a happy ending—occurred during World War II. The British Royal Air Force wanted to attach protective metal plating to the most vulnerable parts of its airplanes, which they identified by tabulating the location of bullet and shrapnel holes on the planes that returned from bombing runs. Most holes were on the wings and rear of the plane, suggesting that this was where plating was needed. However, Abraham Wald, a Jew who had fled the Nazis, recognized the survivorship bias. Few returning planes had holes in the cockpit and fuel tanks because planes hit there had been shot down. The holes in the wings and rear of returning planes were actually evidence that these locations are relatively unimportant. Following Wald's advice, the Air Force put the plates in locations with no holes, which allowed more planes to return safely to Britain. Wald's recognition of survivorship bias helped win the war.

Investors might think that the sleight-of-hand vanishing act for unsuccessful mutual funds is harmless. Poor-performing funds should disappear, and salespeople should show the best funds. But imagine that, instead of paying extravagant salaries to human stock pickers, funds paid monkeys peanuts to pick stocks by flipping coins. A few funds would perform well just because of lucky coin tosses, and investors would not know that the fund's managers are just lucky monkeys.

The unpleasant truth about mutual fund performance is that it is very much like coin flips, in that past performance is of little or no help in predicting future performance—and fund managers are not paid peanuts. As this unpleasant truth has become more widely known, investors have increasingly chosen low-cost index funds which buy a diversified basket of stocks (the index) without paying people to pick stocks. If high-cost funds can't beat the market, then buy the market.

Peddling top-performing funds to gullible investors is similar to a scam where an unscrupulous person sends 1,024 emails with a specific stock prediction: 512 of the emails say, "Starbucks stock will go up tomorrow" and the other 512 emails say the opposite. Whichever way Starbucks goes, the scammer sends follow-up emails to the 512 people who received the correct prediction, again splitting the emails into 256 predictions that "Starbucks stock will go up tomorrow" and 256 saying it will go down. The following day, the 256 people with two correct predictions are divided again, with 128 up-predictions and 128 down-predictions. Eventually, the scammer has 32 fish who have received five correct predictions and may be willing to pay serious money for more of the scammer's predictions.

It's fraudulent, but disturbingly similar to what mutual fund families do when they launch dozens (or hundreds) of funds and promote the winners while burying the losers.

Jay recently saw an infomercial reporting that actively managed funds rarely outperform the S&P 500 index of stock prices. Jay expected the infomercial to recommend investing in an index fund. Nope. They were touting their own trading strategy. Their evidence of its success? Interviews with a handful of satisfied customers! Even if the interviews were real, Jay was sure that there were many unhappy customers who didn't get interviewed. Like unsuccessful mutual funds, they vanished without a trace.

Boaty McBoatface

If people who graduate from college generally get better jobs than people who don't, is it because of the college education or because people who choose to go to college and work hard enough to graduate are generally different from people who don't graduate or don't go to college? If people who took driver's education classes tend to be involved in more traffic accidents than people who didn't, is it because the classes are counter-productive or because of the type of people who feel they need to take a driver's education class? If people who practice the violin more than two hours a day have fewer friends than do people who play in weekly bowling leagues, is it because playing the violin makes a person more introverted? If more people die in bed than anywhere else, is it because beds are dangerous?

These are all examples of the *self-selection bias* that can occur when we collect data by observing what people do. If people choose to do what they are doing, the data may reflect the people rather than the activity. This self-selection bias could be avoided with a controlled experiment in which people are randomly assigned to groups and told what to do. Heads, you have to go to college; tails, you are not allowed to go to college. Fortunately, researchers seldom have the power to make us do things we don't want to do simply because they need experimental data.

In March 2016, the Natural Environment Research Council (NERC) announced a website where the public could provide suggestions for names for a new polar research ship. More than 7,000 people responded. There's a self-selection bias here as well, because pranksters may be especially motivated to participate. The results came in and the winner, by a landslide, was "Boaty McBoatface".

Table 1.4 *NERC name-our-ship campaign*

Top 5 suggestions	Percent of vote
Boaty McBoatface	33.16
Poppy-Mai	10.66
Henry Worsley	4.21
David Attenborough	2.95
It's Bloody Cold Here	2.85

Rather than go with the most popular choice, the NERC announced that the boat would be named after Sir David Attenborough, the famous naturalist. As a consolation prize, a remotely controlled submersible would take the name Boaty McBoatface. In response, an online petition gathered more than 3,000 signatures, urging Sir David Attenborough to change his name to Sir Boaty McBoatface. Good data scientists consider whether their data are tainted by self-selection bias.

Sabermetrics

In the 1970s, Billy Beane was a 6′ 4″, 195-pound star athlete in baseball, basketball, and football at Mt Carmel High School in San Diego. He was big and strong, handsome and charismatic. Beane passed the "eye test" that had traditionally been used to judge a high school athlete's potential.

Stanford offered Beane a combined baseball–football college scholarship, hoping he would play outfield on their baseball team and quarterback on their football team. Beane turned down the scholarship after he was selected in the first round of the Major League Baseball (MLB) draft by the New York Mets. That didn't turn out as well as expected. Over six seasons, he played for four teams, with 315 plate appearances (equal to roughly half a full season), hit three home runs, and batted 0.219. The eye test didn't work.

Ironically, Beane's experience would transform him from a poster-child for the traits valued by traditional scouts ("a five-tool player") to an evangelist for the sabermetrics revolution that emphasizes data, especially overlooked data. For example, a batter who forces a great pitcher to throw lots of pitches is now considered to be a *quality at-bat*, because when pitchers' arms tire, they must be replaced, often by someone who is not as good.

Major League Baseball has always attracted people who like numbers because so many aspects of the game are quantifiable: the number of balls and strikes thrown by a pitcher, the number of base hits made by a batter, the number of errors made by a fielder, the number of runs scored, and the number of games won. Baseball enthusiasts know the last person to bat 0.400 in the regular season (Ted Williams, 0.406 in 1941), the pitcher with the most career wins (Cy Young, 511), and the team that has won the most World Series titles (New York Yankees, 27). Baseball fanatics know many more statistics, some breathtakingly obscure—like the number of stitches on a major league baseball (108).

While working night shifts as a security guard, a baseball fanatic named Bill James studied baseball data and wrote essays on underappreciated aspects of the game. In 1977, he self-published a 68-page compilation of baseball statistics titled *1977 Baseball Abstract: Featuring 18 categories of statistical information that you just can't find anywhere else*. He sold 75 copies. Undeterred, he wrote the *1978 Baseball Abstract: The 2nd annual edition of baseball's most informative and imaginative review*. It sold 250 copies. These modest books with grand titles and meager sales revolutionized our understanding of baseball by championing objective, data-based measurements of player and team performance.

Bill James's pet peeve is that many traditional baseball statistics, while easy to calculate and understand, fundamentally misrepresent a player's importance to a team's success. For example, a player's hitting prowess has customarily been measured by his batting average: the number of base hits divided by the number of times at bat. Sabermetricians consider batting average a poor measure of how much a batter helps his team score runs, which is what really counts—games are won and lost by runs, not hits. Batting averages do not give players credit for getting on-base by walking or being hit by a pitch and do not distinguish among singles, doubles, triples, and home runs, which are of vastly different importance for scoring runs.

One alternative that sabermetricians have developed is on-base plus slugging (OPS), where the on-base average includes hits, walks, and being hit by a pitch, and the slugging average counts the total bases (1 for a single, 2 for a double, 3 for a triple, and 4 for a home run). On-base plus slugging is not a perfect measure, but it is a big improvement over the traditional batting average, and it is now commonly reported in the news and on baseball cards. It is even calculated for Little League players.

One sabermetrics measure of a pitcher's effectiveness is fielding independent pitching (FIP), which counts strikeouts, unintentional walks, hit by pitches, and home runs because these are what a pitcher has control over. When a ball is hit in play (except for a home run), the outcome (hit or out) depends on where the ball is hit and how the fielders handle it, which are beyond the pitcher's control and so are ignored.

A sabermetrics measure of fielding is the range factor (RF), the number of putouts plus assists, divided by the number of games played. An immobile fielder who doesn't make many errors because he doesn't make many plays, is not as valuable as a fielder with great reflexes and range who makes occasional errors, but also makes many outs by fielding balls that lesser fielders don't touch.

Bill James was a data scientist, back when calculations were made by hand using data that were meticulously recorded with pencil and paper. He was a successful data scientist because he understood his objectives, had logical reasons for the novel statistics he created, and tested his theories by seeing how well they correlated with runs scored and games won.

Bill James was initially ignored or laughed at by the baseball establishment—coaches and managers who grew up with traditional baseball statistics and believed in the eye test. That changed with the success of the Oakland Athletics (the A's).

In the late 1980s and early 1990s, the Oakland Athletics were a free-spending, high-achieving team. They won the American League Championship in 1988, 1989, and 1990, and the World Series in 1989. They had all-stars Jose Canseco, Dennis Eckersley, Ricky Henderson, and Mark McGwire, and the highest payroll in baseball in 1991.

The owner, Walter A. Haas, Jr., had hired his son-in-law to be the A's president, and he hired his former law partner, Sandy Alderson, to be the A's general manager. Alderson didn't know much about baseball, but he was smart and demanding. He had gone to Dartmouth College on an ROTC scholarship, was a Marine 2nd Lieutenant in Vietnam, and had a law degree from Harvard. He had enough intelligence and self-confidence to approach baseball from an analytical standpoint, unconstrained by conventional baseball wisdom.

He read Bill James's writings and he was convinced that on-base percentage and slugging percentage not only made sense but, empirically, were better than batting averages for predicting runs scored. He didn't like the bunting, stealing, and hit-and-run plays favored by old-timers because, too often, they gave away precious outs. He explained that, "I figured out that managers do all this shit because it is safe. They don't get criticized for it." John Maynard Keynes once observed that "it is better for reputation to fail conventionally than to succeed unconventionally." As in many other occupations, baseball managers are protected from criticism if they do what other managers do. Alderson believed in his approach and didn't worry about what others thought.

Alderson's analytics were largely ignored when the A's success could be attributed to the all-stars on his team, but Haas died in 1995 and the new owners jettisoned players and slashed the payroll. Now Alderson had to field a competitive team with limited resources, and he used sabermetrics to find undervalued players—players who were cheap but productive. For example, if the Yankees think Bob is better than Juan because Bob has a higher batting average, and Alderson thinks Juan is better than Bob because Juan has a higher OPS, he will try to trade Bob for Juan.

In 1990 Billy Beane had asked Alderson to give him a job as an advance scout to prepare reports on the A's opponents. Alderson hired him, later explaining that "I didn't think there was much risk in making him an advance scout, because I didn't think an advance scout did anything." Soon Beane was reading Bill James and being mentored by Alderson in the science of identifying undervalued players.

Alderson left the A's in 1997 and Beane became the general manager, going all-in on using sabermetrics. In 2002, despite having a $40 million payroll compared with the Yankee's $125 million and 28 other teams that averaged $70 million, the A's won 20 consecutive games on their way to winning the American League West, a season immortalized in Michael Lewis's book *Moneyball*.

After the success of the 2002 A's demonstrated the value of sabermetrics, Oakland's advantage has dissipated and perhaps disappeared, now that many of the statistics that sabermetricians developed and championed have become commonplace. A player's OBS, for example, used to be undervalued, but now it is fairly valued, which has taken away the A's competitive advantage.

In 2002 a group led by a billionaire investor named John Henry bought the Boston Red Sox, and Henry hired Bill James to help break the Curse of the Bambino—the Red Sox having gone more than 80 years without winning the World Series, presumably because the Red Sox owner sold the team's best player, Babe Ruth ("the Bambino") to the New York Yankees in 1920 for $125,000 cash and a $300,000 loan to finance a Broadway musical starring the owner's girlfriend. The Red Sox broke the curse in 2004 and won the World Series again in 2007, 2013, and 2018.

In 2015, ESPN's Ben Baumer categorized each MLB baseball team according to their use of data analytics and labeled them "all-in", "believers", "one foot in", "skeptics", or "nonbelievers." Nine of the ten teams that made the playoffs that year were in one of the first two categories. Every team now has sabermetricians, using far more sophisticated metrics than OBS, FIP, and RF. One appealing overall measure of a player's value is wins above replacement (WAR), which is an estimate of how many wins a player contributes to his team compared to an easily acquired replacement player—someone who plays the same position in the minor leagues and is not quite good enough to play in the majors.

A WAR of zero means that the player performed no better than an inexpensive replacement. As a baseball player, Billy Beane's lifetime WAR was –1.6. Even though he passed the eye test, he was not good enough to play in the majors. As a baseball manager, Beane's WAR is extremely high, whether measured by the number of games the A's won because of him, or by the way his approach revolutionized baseball.

Data science has now spread to virtually all sports. Good data scientists begin with a clear understanding of the game, and then use data to test and quantify their theories. Any data clown can rank players by historically popular metrics, but recognizing the shortcomings of common statistics and creating better ones is the hallmark of an expert.

Good data

Good data scientists consider the reliability of the data, while data clowns don't. It's also important to know if there are unreported "silent data." If something is surprising about top-ranked groups, ask to see the bottom-ranked groups. Consider the possibility of survivorship bias and self-selection bias. Incomplete, inaccurate, or unreliable data can make fools out of anyone.

The First Pitfall of Data Science is:

Using Bad Data

Putting Data Before Theory

> "Our software found the oddest thing. People with long last names
> love our product!"
> "So let's focus our sales team on people with long last names."
> "But it doesn't make any sense."
> "Up is up."

An Internet marketing company (TryAnything) gave website visitors
access to some "hot deals" if they filled out a questionnaire. Their data
scientists then ransacked the data, looking correlations between the questionnaire answers and hot-deal purchases. They found something truly
surprising—people with more than nine letters in their last name were
more likely to buy something. No one had a compelling explanation, but
TryAnything's management didn't need an explanation. They just said "up
is up," which is a common catchphrase used by people who believe that
data speak for themselves. Who needs a reason? If it works, use it.

TryAnything identified everyone in its database who had a long last
name and had its salespeople make personal telephone calls to each. No
robocalls. Real calls by real people who were paid real money to make the
calls. It was a colossal flop. The calls were a money pit and were abandoned after a few months.

What went wrong? It's called data mining, and good data scientists are
wary of data mining.

The 9 Pitfalls of Data Science. Gary Smith and Jay Cordes. Oxford University Press (2019).
© Gary Smith and Jay Cordes 2019. DOI: 10.1093/oso/9780198844396.001.0001

Data mining

The traditional statistical analysis of data follows what has come to be known as the *scientific method* that replaced superstition with scientific knowledge. For example, scientists know that aspirin inhibits the ability of blood platelets to form clots that stop blood flowing from damaged blood vessels. Scientists also know that heart attacks and strokes can be triggered by blood clots (heart attacks occur when blood vessels leading to the heart are clotted; strokes occur when blood vessels leading to the brain are clotted.)

A natural research hypothesis is whether regular doses of aspirin can reduce the chances of heart attacks and strokes for people who are at risk. Following the scientific method, researchers would test this theory by gathering data, ideally through a controlled experiment.

That was precisely what was done in the 1980s when 22,000 male doctors volunteered to participate in an experiment. Half the doctors were given an aspirin tablet every other day; the other half took a placebo— something that looked and tasted like aspirin, but was an inert substance. The test was double-blind in that neither the volunteers nor the researchers knew which volunteers were taking aspirin.

After five years, an outside committee that was monitoring the results found that 18 of the doctors taking placebos had fatal heart attacks, compared to only 5 of the doctors taking aspirin. Nonfatal heart attacks were experienced by 171 doctors taking placebos and by 99 doctors taking aspirin. Both results were highly significant. If aspirin were no more effective than the placebo, there was only a 0.0067 chance of this large a difference in fatal heart attacks, and a 0.0000105 chance of this big a difference in nonfatal heart attacks. The American Heart Association now recommends that, "People at high risk of heart attack should take a daily low-dose of aspirin (if told to by their healthcare provider) and that heart attack survivors regularly take low-dose aspirin."

These researchers used data to test a theory. Data mining goes in the other direction, analyzing data without being motivated or encumbered by theories. When an interesting pattern is found, the researcher may argue that we don't need an explanation because data are sufficient. Up is up.

In addition to those who believe that reasons are not needed, some believe that data can be used for *knowledge discovery*, a data-driven revelation of new, heretofore, unknown relationships. In our heart-attack example, a data miner might compile a database of, say, 1,000 people who suffered heart attacks and record everything we know about them, including income, hair color, eye color, medical history, exercise habits, and dietary habits. Then data-mining software is used to identify the five personal characteristics that are most common among these victims in comparison with the general population. These five characteristics might turn out to be low banana consumption, above-average income, living near a post office, green eyes, and using the word *fantastic* frequently on Facebook.

The data miner might conclude that no-bananas, above-average income, post offices, green eyes, and Facebook *fantastic*s are unhealthy, and concoct some fanciful stories to explain these correlations. Or the data miner might believe that no explanation is needed. As two data scientists recently put it, data mining is a quest "to reveal hidden patterns and secret correlations."

The evolution of data mining

Decades ago, being called a "data miner" was an insult comparable to being accused of plagiarism. If someone presented research that seemed implausible or too good to be true (for example, a near perfect correlation), a rude retort was "data mining!" Today people boast of being data miners. What was once considered a sin has become a virtue.

In 2008, Chris Anderson, editor in chief of *Wired*, wrote an article with the provocative title, "The End of Theory: The data deluge makes the scientific method obsolete." Anderson argued that

The new availability of huge amounts of data, along with the statistical tools to crunch these numbers, offers a whole new way of understanding the world. Correlation supersedes causation, and science can advance even without coherent models, unified theories, or really any mechanistic explanation at all.

A 2015 article in *The Economist* argued that macroeconomists (who study unemployment, inflation, and the like) should abandon the scientific method and, instead, become data miners:

[Macroeconomists] should tone down the theorizing. Macroeconomists are puritans, creating theoretical models before testing them against data. The new breed ignore the white board, chucking numbers together and letting computers spot the patterns.

The Economist is a great magazine, but this was not great journalism.

Sometimes the term *data mining* is applied more broadly, to encompass such useful and unobjectionable activities as Internet search engines and DNA database queries in murder investigations. We use *data mining* specifically to describe the practice of using data to discover statistical patterns that are then used to predict behavior. For example, data miners search for statistical relationships that might be used to predict car purchases, loan defaults, illnesses, or changes in stock prices.

Data mining is known by many names, including data exploration, data-driven discovery, knowledge extraction, information harvesting, data dredging, fishing expeditions, cherry picking, data snooping, and p-hacking—all reflecting the core idea that data come before theory. The p-hacking nickname comes from the fact that a pattern in the data is considered statistically significant if there is a low probability (a low p-value) that it would occur by chance. A data miner has a high probability of reporting a low p-value, so he is a p-hacker.

The Texas Sharpshooter Fallacy

The endemic problem with data mining is nicely summarized by the two Texas Sharpshooter Fallacies:

1 A wanna-be cowboy paints hundreds of targets on a wall. When he fires his gun, he hits one of the targets, erases all evidence of the targets he missed, and proudly points to the target he hit as proof of his marksmanship. This evidence is worthless because, with so many targets, he was bound to hit one. In data science, this is analogous to doing hundreds (or thousands or millions) of tests and reporting only the ones with low p-values. In a study of the stock market, for example, a data scientist might look at stock returns on each day of the year over the past 100 years, and report that November 22 is a good day to buy stocks.

2 Alternatively, the wanna-be cowboy shoots a bullet at a blank wall, and then draws a target around the bullet hole. This evidence of his marksmanship is worthless because the target was drawn after there was a bullet hole to draw the target around. In data science, this is equivalent to discovering a statistical pattern and concocting a theory to match the pattern. A stock market observer might remember that some especially good years in the stock market had been preceded by snowy Christmas Eves in Boston, and make up some fanciful explanation. (Yes, there really is a Boston snow (B.S.) indicator.)

Texas Sharpshooter Fallacy #2 is also called the Feynman Trap, a reference to Nobel Laureate Richard Feynman. In one of his Caltech classes, Feynman asked what the probability was that, if they walked outside the classroom, the license plate number on the nearest car in the parking lot would be 2AXY47. His students thought this was an easy problem (Caltech students are very smart). They assumed that these three numbers and three letters were randomly selected from all possible numbers and letters and calculated the probability to be less than 1 in 17 million. After they reported their answer, Feynman announced that the probability was actually 1 because he had seen that license plate before he walked into the classroom. A seemingly unlikely event is not at all unlikely if it has already happened.

Sports Illustrated reported that 2015 was an incredible sports year (the "Best Year Ever") because it made "a mockery of the odds." They looked at the five remarkable events below and calculated the probability of each, based on the betting odds before the event happened:

U.S. women's soccer team winning World Cup:	1 in 35
American Pharoah winning Triple Crown in horse racing:	1 in 9.9
Kentucky men's college basketball team being undefeated:	1 in 17.7
Serena Williams winning tennis' grand slam:	1 in 70.4
Jordan Spieth winning 2 of 4 golf majors:	1 in 3,118.5

A statistician calculated the probability of all five events happening to be 1 in 522,675.

He fell into the Feynman Trap. If *Sports Illustrated* (or the statistician) had predicted these five events before they happened, that would have been incredible. Not so incredible, predicting outcomes *after* they occur. There were thousands of sporting contests in 2015 and the probability of identifying several unlikely outcomes *after* they occurred is one.

Cell phones don't cause cancer

Sometimes people get notions that have no theoretical or empirical basis. For example, the idea that cell phones cause cancer comes from an unreasonable fear of anything with the word "radiation" in it. Not all radiation is equal. For the question of whether radiation causes cancer, Figure 2.1 shows that there is an important distinction between ionizing and nonionizing radiation.

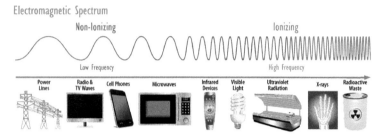

Figure 2.1 Ionizing versus non-ionizing radiation. Credit: National Institute of Environmental Health Sciences

The higher the frequency, the higher the energy, and the deeper the radiation can penetrate a human body. That's why ultraviolet light, at the low end of the spectrum for ionizing radiation, can cause only skin cancer, while X-rays can penetrate the skin and cause cancer in internal organs. There is no known mechanism by which nonionizing radiation (which includes Wi-Fi, radios, and cell phones) has enough energy to damage DNA and cause cancer.

There may be a currently unknown mechanism by which cell phones cause cancer but, if so, that link should be supported by data showing that cancer rates have increased with cell-phone usage. For example, Figure 2.2 shows that the male lung cancer death rate in the United States has risen and fallen about 20 years after similar trends in per capita cigarette consumption, suggesting that there might be a relationship between tobacco use and lung cancer.

The data for cell-phone usage and brain cancer is very different. Figure 2.3 shows that while the number of U.S. cell-phone subscribers has exploded, the incidence of brain cancer has declined.

Historical relationships are only suggestive since it is easy to find coincidental correlations that are meaningless. For example, Figure 2.4 shows that there is a nonsensical correlation between the length of the winning word in the Scripps National Spelling Bee and the number of Americans killed each year by venomous spiders.

In the case of cigarettes and lung cancer, scientists have shown that many chemicals found in cigarettes cause DNA damage, and there is overwhelming epidemiologic and experimental evidence that cigarette smoking can cause cancer, thereby explaining the historical relationship in Figure 2.2.

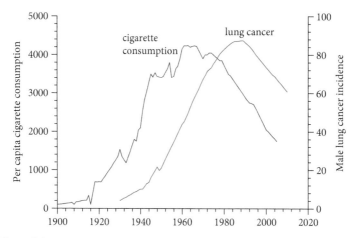

Figure 2.2 Historical relationship between cigarette smoking and lung cancer

In the case of cell-phone usage and brain cancer, however, the historical evidence in Figure 2.3 does *not* suggest that cell phones cause brain cancer. For decades, brain cancer rates have been *falling* while cell-phone usage has been skyrocketing, which is consistent with the scientific evidence that nonionizing radiation does not cause cancer.

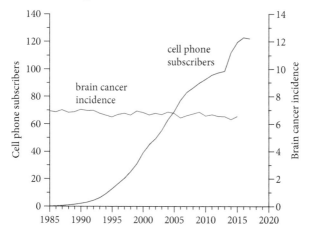

Figure 2.3 U.S. cell phone subscribers (per 100 population) and brain cancer (age-adjusted per 100,000 population)

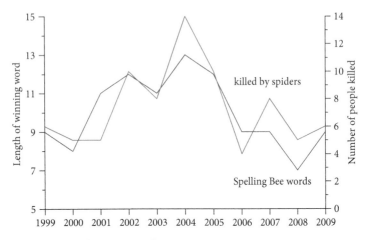

Figure 2.4 Be careful what you spell

Nonetheless, a Texas Sharpshooter who persists in looking for correlations is sure to find something, just like the person who discovered the correlation between spelling-bee words and spider-bite deaths. A study called Interphone looked at 14,078 people in 13 countries, of which 6420 people had some kind of brain cancer and 7658 did not. These were not experimental data and the researchers had to rely on the subjects' recollection of their cell-phone usage, which might be inaccurate and biased. For example, some people with brain cancer might blame it on their cell phones by exaggerating the amount of time spent using cell phones.

The Interphone study found that, overall, there was no evidence of any risk from cell-phone usage. There was a weak association between phone usage and a rare type of cancer for people with the most cell-phone usage; however, people who did not use cell phones at all had substantially more brain tumors than did light users—suggesting that cell phones are a defense against brain cancer!

These two contradictory blips in the data are the kind of fluky things that Texas Sharpshooters find when they look for patterns in many places. It is telling that there was no consistent pattern of tumors either rising or declining as we move through the deciles of cell-phone usage—just two peculiar and conflicting blips at the extremes. The report noted the blips and concluded that, "biases and errors limit the strength of the conclusions we can draw from these analyses and prevent a causal interpretation." A discovered correlation has to be very convincing if it is to be taken seriously, and these two blips are not convincing.

Good data scientists avoid data mining and demand compelling evidence from Texas Sharpshooters.

Google Flu

In 2011, Google created an artificial intelligence program called Google Flu that used search queries to predict flu outbreaks. They boasted that, "We can accurately estimate the current level of weekly influenza activity in each region of the United States, with a reporting lag of about one day." They said that their model was 97.5 percent accurate, in that the correlation between the model's predictions and the actual number of flu cases was 0.975. An MIT professor said that, "This seems like a really clever way of using data that is created unintentionally by the users of Google to see patterns in the world that would otherwise be invisible. I think we are just scratching the surface of what's possible with collective intelligence."

How did Google do it? Google's data-mining program looked at 50 million search queries and identified the 45 queries that were the most closely correlated with the incidence of flu. It was pure-and-simple data mining and a terrific example of the Feynman Trap.

A valid study would use medical experts to specify a list of relevant query phrases in advance, and then see if there was an uptick in these queries shortly before or during flu outbreaks. Instead, Google's data scientists had no control over Google Flu's selection of the optimal search terms. The program was on its own, with no way of telling whether the search queries it found were sensible or nonsense.

Since flu outbreaks are highly seasonal, Google Flu may have been mostly a winter detector that chose seasonal search terms, like *Christmas*, *winter vacation*, and *Valentine's Day*. With 50 million search terms, there are also bound to be some queries, like *Jeff Goldblum* and *Steve Carell*, that happened, by chance alone, to be highly correlated historically with flu cases, but useless for predicting future outbreaks.

When it went beyond fitting historical data and began making real predictions, Google Flu was far less accurate than a simple model that predicted that the number of flu cases tomorrow will be the same as the number today. After issuing its report, Google Flu overestimated the number of flu cases for 100 of the next 108 weeks, by an average of nearly 100 percent.

Google Flu no longer makes flu predictions.

Target's pregnancy predictor

Since its rebranding in the 1990s as a store for "younger, image conscious shoppers," Target has grown to become the second largest discount retailer in the United States (the largest is Walmart, with its focus on "always low prices"). One of Target's strengths has been its analysis of customer data. Target gives each customer a unique guest ID number and records everything the customer buys; responses to Target emails, web site visits, and coupons; and information purchased from data venders.

For example, Target's data scientists used these data for the women in Target's baby registry to develop a model for predicting whether a Target shopper is pregnant. If Target's data scientists had been data clowns, they might have acted like Texas Sharpshooters by ransacking Target's vast database for statistical correlations and discovered some useless coincidences, like an uptick in the purchases of toothpaste and men's socks during the first trimester.

Instead, the data scientists thought about how someone might guess that a woman is pregnant just by watching what she buys. They focused on plausible changes in buying habits and were able to use data for 25 reasonable products to estimate the probability that a woman was pregnant and to predict her due date. They found that pregnant women tend to buy more dietary supplements, like calcium, during the first half of their pregnancies; switch to unscented soap and lotion during their second trimester; and stock up on cotton balls and washcloths during their third trimester.

Target sent baby coupons and special offers to women who their model identified as having a high probability of being pregnant. It worked great until an angry father demanded to talk to a manager: "My daughter got this in the mail! She's still in high school, and you're sending her coupons for baby clothes and cribs? Are you trying to encourage her to get pregnant?"

A few days later, when the manager called him to apologize, the father's tone was different: "I had a talk with my daughter. It turns out there's been some activities in my house I haven't been completely aware of. She's due in August. I owe you an apology."

Target's model worked because it wasn't data mining. Target compared the shopping habits of women who were identified as pregnant with women who weren't, and they only considered products that made sense.

A data-mining approach would have discovered hundreds of characteristics the women shared in common, like purchases of socks and cat food. The results would have been an unintelligible goulash of apples, oranges, and potatoes. Target didn't do that, but other companies do.

Fighting crime with Facebook

A data-savvy police department data mined the Facebook accounts of local residents to see if surges and slumps in the use of certain words might be helpful in predicting criminal activity. They identified the 100 most popular nouns, 50 most popular adjectives, and 50 most popular adverbs in the English language. Then they collected daily data for 10 weeks on the frequency with which each of these 200 words were used in status updates on Facebook and the number of burglaries committed the next day. All the data were scaled to equal 100 at the start of the study, thus a value of 101 means 1 percent more than initially, and 99 means 1 percent less.

They found that the two most helpful words for predicting burglaries were *day* and *most*. Figure 2.5 shows the close correspondence between actual burglaries and those predicted by their computer model using these two words. The correlation between predicted and actual burglaries was 0.96.

The number of burglaries more than doubled during this 10-week period, and their Facebook model accurately predicted this increase and most of these smaller ups and downs. We might leave it at that. Up is up, so we don't need any logical explanation of why these two specific words are so useful in predicting burglaries. Or we might make up explanations. Perhaps the word *day* shows up because people are communicating about when to commit the burglary. The word *most* might be used to communicate information about the target, tools, or something else. Or perhaps *day* and *most* are code words used by local burglars, and the police have broken the code.

Do we really need a logical explanation or is it enough to say that we used data science to find a way to anticipate the ebb and flow of criminal activity? Actually, it does matter. Statistical patterns that are unearthed by data mining and have no logical explanation are often worthless.

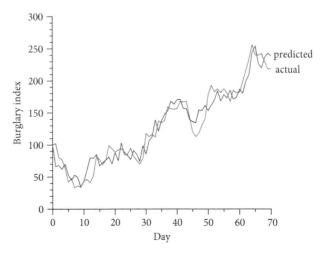

Figure 2.5 Using Facebook to anticipate crime

Okay, time for a confession: no police department did this. We created all the data using a computer's random number generator. The burglary data and the data for each word started at 100 and then followed a random walk by going up or down each imaginary day, using random, independent draws from a normal distribution. There is no real relationship between the words and the "burglaries," because all the data were generated independently, but there were several purely coincidental correlations between the made-up word data and the made-up burglary data. The two words with the strongest correlation just happened to have the randomly assigned labels *day* and *most*.

That is the intended lesson. If a data clown sifts through enough useless data, some strong, but totally coincidental, statistical patterns are bound to be discovered. If useless data can make accurate predictions, then accurate predictions are not enough to demonstrate that a data-mined model is useful for making predictions with fresh data.

Figure 2.6 shows how this Facebook-word model worked over the next 10 weeks. The predicted number of burglaries trended down, while the actual number of burglaries trended up. The day-to-day fluctuations in predicted and actual burglaries were sometimes in the same direction, other times the opposite. Overall, the correlation between the model's

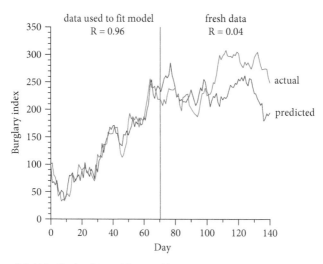

Figure 2.6 Using Facebook to anticipate nothing

predictions and actual burglaries was 0.04. The model was essentially use-less—which is not a surprise because all the fluctuations in burglaries and words were independently generated random numbers.

We are not saying that Facebook data are useless or, more generally, that statistical relationships are worthless. What we are saying is that data mining may not be knowledge discovery, but noise discovery. There needs to be more than mere statistical patterns—there should be theory or con-trolled experiments or both. It is far too tempting to say, "up is up," and believe that every discovered pattern is meaningful, and then be disap-pointed when a pattern vanishes.

This hypothetical Facebook/burglary model is very similar to Google Flu's use of search queries to predict flu outbreaks. In later chapters, you will see that others have found similar correlations and thought they were meaningful—including using Facebook words to price auto insurance and using Twitter words to predict stock prices.

The set-aside solution

A data-mined model is likely to be useless, and this uselessness is likely to be revealed when the model is used to make predictions with new data.

So, it might be thought that a surefire way to distinguish between a useful model and a useless one is to separate the data into two parts—one part for choosing the model and one part for testing the model. The original data are called *training data*, while the set-aside data are called *validation data*. An alternative set of labels is *in-sample* (the data used to discover the model) and *out-of-sample* (the data used to validate the model).

It is certainly good data science practice to set aside data to test models. However, suppose that we data mine lots of useless models, and test them all on set-aside data. Just as some useless models are certain to fit the original data, some, by luck alone, are certain to fit the set-aside data too. Finding a model that fits both the original data and the set-aside data is just another form of data mining. Instead of discovering a model that fits half the data, we discover a model that fits all the data. That makes the problem less likely, but doesn't solve it.

To demonstrate this, let's look again at the 200 words that were used to predict daily fluctuations in burglary data. There are 19,900 possible two-word models: words 1 and 2, words 1 and 3, and so on. We fit all 19,900 of these two-word models to the 10 weeks of in-sample data and then saw how well they did for the 10 weeks of out-of-sample data. The results are in Figure 2.7.

For the 10 weeks of data used to estimate the models, the correlation between the predicted and actual values cannot be less than zero, because the best-fit model can always ignore the word variables completely and have a correlation of zero. The average correlation for the in-sample models was 0.74.

For the 10 weeks of out-of-sample data that were set aside to test the models, the correlation is equally likely to be positive or negative because the words are, after all, random numbers that have nothing at all to do with the burglary data. We expect the average correlation to be close to zero. For these particular data, the average out-of-sample correlation happened to be –0.04.

Nonetheless, some out-of-sample correlations were, by chance, strongly positive and others were strongly negative. Figure 2.7 shows that several models fit the in-sample data well and also do well out-of-sample, sometimes even better than in-sample. That is the nature of chance, and these are chance variables.

In-sample, where the models are fit to the data, hundreds of models have a correlation above 0.90. Out-of-sample, where the models are

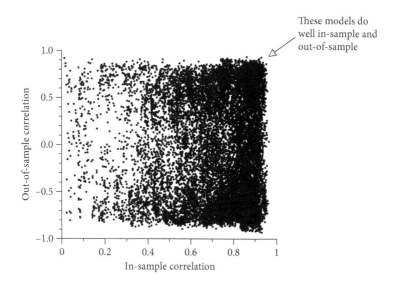

These models do well in-sample and out-of-sample

Figure 2.7 In-sample and out-of-sample fit for 19,900 two-variable models

accurate only by chance, there are nonetheless dozens of models with a correlation above 0.90. Five models have a correlation above 0.90 both in-sample and out-of-sample. These five models pass the validation test with flying colors even though it was pure luck. They are still useless for predicting burglary rates in other weeks.

One model, using words 107 (*thing*) and 173 (*kid*), had a correlation of 0.92 in-sample and an even higher correlation (0.93) out-of-sample! If we didn't know better, we might think that we had discovered something important. But, of course, we didn't. All we really discovered is that it is always possible to find models that do well in-sample and out-of-sample, even if the data are just random noise.

Models should be tested with set-aside data, but set-aside data are not a cure for energetic data miners. Good data scientists are wary of data mining.

Data mining heart-attack data

Let's look again at the aspirin/heart-attack example, in which a controlled experiment concluded that regular doses of aspirin could reduce the risk of a heart attack. A data miner would approach the study of heart attacks quite differently. A data miner might collect lots of data on heart-attack victims and look for traits and behaviors they had in common. Earlier, we speculated that the five personal characteristics that are most common among these victims might turn out to be low banana consumption, above-average income, living near a post office, green eyes, and often using the word *fantastic* on Facebook.

Perhaps you thought we were joking. We were, of course, but many jokes are based on truths. To demonstrate this, we made up some more completely random data. First, we created 1000 imaginary heart-attack victims. Then we looked at 100 household expenditure categories; for example, cereal, prescription drugs, and outdoor tools. For each of the 1000 imaginary heart-attack victims, we created fictitious data on spending in each of the 100 categories by making random draws from a normal distribution with a mean of zero. Positive values mean that this person's spending in this category was above average in comparison to the general population, and negative values mean below-average spending.

There is absolutely no systematic relationship between heart attacks and spending since these data were created independently. Although each random spending number was equally likely to be above average or below average, spending in some categories will happen, by chance, to be significantly above or below average.

That is exactly what happened. There were five categories in which average spending was significantly high or low. The most statistically significant item was above-average spending on fish. These heart-attack victims also spent more on men's footwear and less on pork, cheese, and cleaning products.

Some data miners might say, "up is up." The data show that these are statistically significant risk factors for heart attacks, and we don't need reasons. Other data miners might celebrate their knowledge discovery and make up theories about why eating fish and buying men's footwear, while skimping on pork, cheese, and cleaning products, helps prevent heart attacks. You might even be thinking up theories yourself—a fish diet is healthy, people who buy a lot of men's footwear exercise a lot, and so on.

It is certainly tempting (and easy) to think up explanations after the fact. But, remember, these are *completely* made-up data. Product 27 happened, by chance, to have an unusually high spending number, and product 27 happened, by chance, to be given the label, "fish." These are not real people and these are not real spending data, and that is precisely our point.

A statistical relationship is considered statistically significant if there is only a 5 percent (or lower) chance that it would happen by luck alone. Fair enough, but Texas Sharpshooter Fallacy #1 reminds us that, even if we look at random data, we can expect 5 percent of our results to be statistically significant. The more relationships we test, the more statistically significant relationships we can expect to find. One out of every 20; 5 of every 100; 50 of every 1000.

Not only that, but even if some of our tests involve true relationships, the more tests we do, the greater the probability of finding something useless. Imagine that there are 5 important risk-factors for heart attacks. A data miner who happens to look at these 5 true factors and 20 worthless factors might find 6 statistically significant relationships—5 real and 1 phony. A more ambitious data miner who looked at these 5 true factors and 100 worthless factors might find 10 statistically significant relationships—5 real and 5 phony. An investigation of 5 true factors and 1000 worthless factors might find 5 real and 50 phony. The more factors that are investigated, the greater the chances that a statistically significant factor is phony.

It is even worse than these dismal odds because there is no guarantee that a data miner who blithely ignores theory will even consider the five real factors. And even if real factors are included in the analysis, there is no guarantee that they will turn out to be statistically significant. In any given set of data, just as some meaningless factors might turn out, by luck alone, to be statistically significant, so some meaningful factors might turn out, by luck alone, not to be statistically significant.

Here, we tested 100 worthless theories about heart attacks and, no surprise, found 5 statistically significant relationships. We know these 5 theories are worthless, because we know that the data are just random numbers. When data miners ransack data, they do not know whether the statistical relationships are real or phony because, remember, they do not want to restrict their analysis by thinking about theories before they look at the data. After they get their results, they can always invent explanations for whatever they find (fish is good for us, runners buy footwear).

What we can be sure of is that data mining can be expected to discover patterns that are statistically significant, but worthless, and the more data that are mined, the great the probability that the discovered relationships are worthless.

Tweet, tweet

Gary recently received a prospectus for a hedge fund (ThinkNot) that boasted:

Our fully automated portfolio is run using computer algorithms. All trading is conducted through complex computerized systems, eliminating any subjectivity of the manager.

This fund's decisions to buy and sell stocks are based solely on a data-mining algorithm with no consideration whatsoever of whether the trades make sense—because that would be "subjective." What they call *subjective*, we call *expertise*.

Their algorithm might consider stock prices, interest rates, the unemployment rate, the number of times "happy" is mentioned in tweets, sales of yellow paint, aspirin consumption, and dozens of other variables. The output might be a decision to buy 100 shares of Apple stock. The fund turns its investment decisions over to a computer because the fund's managers think that computers are smarter than humans. Up is up.

Did you think that we were joking when we said that the ThinkNot data-mining algorithm might look at the number of times "happy" is mentioned in tweets? In 2011, a team of researchers reported that their data-mining analysis of nearly 10 million Twitter tweets during the period February to December 2008 found that an upswing in "calm" words predicted an increase in the Dow Jones average up to six days later. The lead author said he had no explanation, but there is an obvious one: they were Texas Sharpshooters.

The Voleon Group

In 2008 Michael Kharitonov and Jon McAuliffe, with PhDs in computer science and statistics, respectively, launched an algorithmic trading company they named Voleon. A 2010 *Wired* article was enthusiastic, noting that their trading algorithms operate on their own, with no human supervision:

McAuliffe and Kharitonov say that they don't even know what their bots are looking for or how they reach their conclusions. "What we say is 'Here's a bunch of data. Extract the signal from the noise,'" Kharitonov says. "We don't know what that signal is going to be like."

Voleon's algorithms reportedly sift through all kinds of data, including satellite images, credit-card receipts, and social-media language, looking for patterns related to stock prices. Part of their machine learning involved an analysis of terabytes of data on every price change of every stock over a 15-year period.

Ten years later, in December 2017, a *Wall Street Journal* article reported on their difficulties. Voleon's annual return had been a distinctly mediocre 10.5 percent, slightly worse than the 10.7 percent return for the S&P 500. Kharitonov said,

Most of the things we've tried have failed…The idea that we could just take the machine-learning techniques in speech recognition and computer vision to generate better forecasts just didn't work.

The article attributed Voleon's struggles to the volatile nature of markets:

The basic problem they faced was that markets are so chaotic. Machine-learning systems have been best applied so far to situations where patterns are more of a repeating nature, and thus easier to discern, such as in playing the ancient game of Go or even guiding a driverless car. The financial markets are "noisier"—continually being affected by new events, the relationships among which are frequently shifting.

This explanation is only partly true and misses the bigger point. It *is* easier for algorithms to deal with board games with fixed rules. However, the problem with stock-trading algorithms is not just that markets are buffeted by unexpected events, but that the statistical patterns the algorithms discover are often fortuitous and worthless. If an algorithm finds a historical correlation between stock prices and the use of calm words on Twitter, and the pattern disappears when it is used to buy and sell stocks, it is not because the world has changed, but because there never was a real relationship—just a transitory coincidental correlation.

England loves teal

An Internet marketer (QuickStop) owned more than a million domain names that it used to lure inadvertent web traffic. For a hypothetical example, someone who wanted to go to the checkreorderexpress.com to reorder checks, but mistakenly typed checkorderexpress.com was sent to a QuickStop landing page (a *lander*), where the company peddled products (including checks).

A QuickStop senior manager thought the company might boost its revenue if it changed its traditional blue landing-page color to yellow, red, or teal. The company's data analysts set up four versions of the landing page and, after several weeks of tests, concluded that none of the alternative colors led to a significant increase in revenue. The senior manager was disappointed and suggested that they might find some differences if they separated the data by country.

Sure enough, when the data were broken down by country, they found one nation where there was a statistically significant result: England loves teal. Great Britain is an island and maybe the teal color reminded the Englanders of the waters surrounding Great Britain? The manager said that it didn't really matter why; the important point is that the English spend more when the landing page is teal. Up is up.

The manager wanted the teal layout to be shown to all English web traffic, but a QuickStop data scientist blew an imaginary whistle and called a "data foul." This was pure data mining, and Texas Sharpshooter Fallacy #1. QuickStop hadn't initiated the experiment with the thought that England might love teal. By looking at three alternative colors for 100 or so countries, they virtually guaranteed that they would find a revenue increase for some color for some country—that they would hit one bullseye out of hundreds of targets. The more colors and the more countries, the more likely it is that—by chance alone—they would find something significant.

The data scientist insisted that they continue the test and gather additional data. He wasn't surprised when the teal effect disappeared. In fact, revenue from England was significantly lower with the teal landing page than with the company's original blue color.

The same was true in the TryAnything example at the beginning of this chapter. TryAnything's questionnaires had dozen of entries and some answers—by chance alone—were bound to be correlated with product sales. But QuickStop's data scientist was not working for TryAnything, so the company was not hit with a data foul. QuickStop wasted money because it fell for a Texas Sharpshooter Fallacy.

Simpson's Paradox

On another occasion, the QuickStop analysts were stumped by a particularly vexing puzzle. Some landing pages have a "1-click" format, with advertisements appearing on the first page. Other sites have a "2-click" format, with keywords on the first page that, if clicked, send the user to a second page with ads targeted to the keyword. The 2-click format requires more user effort, but the targeted ads are more likely to be effective.

The QuickStop analysts compared the revenue per user for the two formats, and were baffled by data similar to the numbers in Tables 2.1 and 2.2. Table 2.1 shows that RPM (revenue per thousand users) was higher with the 2-click format ($16.40 versus $10.00), which led them to believe that they could increase revenue by swapping out all of the 1-clicks for 2-clicks.

However, when they separated the data into United States versus International (all non-US countries) in Table 2.2, they found that 1-click landers had much higher revenue for both!

Everyone was perplexed. How could one format be superior inside and outside the United States, yet inferior overall? Some people checked the math; others checked their email. Eventually, a data scientist remembered something she had seen several years ago in a statistics class: Simpson's Paradox.

Simpson's Paradox occurs when aggregated and disaggregated data lead to contradictory conclusions. Here, the aggregated data indicate that 2-click landers are superior, but when the data are disaggregated into United States and international, 1-click landers are superior.

Figure 2.8 shows why this happens. For both U.S. and international users, RPM is lower for 2-click landers than for 1-click landers. The averages tell a different story—a misleading story. For 1-click landers, 75 percent of the users were international, making the average RPM value closer to the RPM for international users. For 2-click landers, 80 percent of the

Table 2.1 *Revenue, users, and revenue per thousand users (RPM)*

1-click landers			2-click landers		
Revenue	Users	RPM	Revenue	Users	RPM
$4,000,000	400,000,000	$10.00	$4,100,000	250,000,000	$16.40

Table 2.2 *United States and international users considered separately*

	1-click landers				2-click landers		
	Revenue	Users	RPM	Revenue	Users	RPM
United States	$2,500,000	100,000,000	$25.00	$4,000,000	200,000,000	$20.00
International	$1,500,000	300,000,000	$5.00	$100,000	50,000,000	$2.00
Total	$4,000,000	400,000,000	$10.00	$4,100,000	250,000,000	$16.40

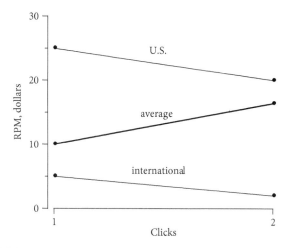

Figure 2.8 Simpson's paradox for 1-click and 2-click landers

users were U.S. users, making the average RPM value closer to the RPM for U.S. users. Because of this compositional change, the average RPM is higher for 2-click landers than for 1-click landers.

The type of user—U.S. or international—is a *confounding* variable in that RPM is related not only to the type of lander, but also to the country of the user. If we take this confounding factor into account by separating the data into U.S. and international users we see that the 1-click RPM is higher for both types of user. Quickstop decided to quickly stop data mining and stick with experiments.

Confounding variables are our enemies if they prevent us from understanding the true nature of a relationship. Controlled experiments are our friends, because they allow us to identify what really matters and what doesn't.

Step away from stepwise

Stepwise regression is a very efficient data-mining technique. Instead of considering all possible combinations of variables for, say, predicting stock returns, stepwise regression adds variables, one at a time, by selecting the variable with the lowest *p*-value at each step. This selection procedure is simple and fast, and many researchers working with big data have succumbed to the appeal of stepwise regression.

The reality, however, is that the larger the number of potential variables, the more likely it is that data mining will select useless variables. Stepwise regression steps—indeed, leaps—into this trap. Its automated rules pay no attention to whether the candidate variables are sensible. It is data without theory. It is data mining on steroids.

We constructed a simulation model for predicting daily stock returns in order to demonstrate this pitfall and how it is exacerbated with large data sets. In each simulation, there were five *true* variables that determined the stock returns and several *nuisance* variables that have no effect on stock returns, but might be coincidentally correlated with them. One hundred observations were used to estimate the stepwise model, and the remaining 100 observations were used to test the model's reliability.

Table 2.3 summarizes the results. With 100 candidate variables, for example, stepwise regression chose an average of 10.71 variables; 60 percent of the variables chosen were nuisance variables. Table 2.3 also compares the in-sample and out-of-sample average prediction errors. (These are absolute values, so that positive and negative errors don't cancel out, like the unfortunate statistician who drowned walking across a river with an average depth of two feet.)

As the number of candidate variables increases, stepwise regression is increasingly likely to choose nuisance variables that are coincidentally related to stock returns in-sample, but useless out-of-sample—causing the in-sample fit to improve, while the out-of-sample fit deteriorates.

Good data scientists say *no* to data mining and step away from stepwise.

Table 2.3 *Stepwise stumbles over big data*

Number of candidates	Average number of variables chosen	Probability of being a nuisance variable	Average error in-sample	Average error out-of-sample
10	4.74	0.06	15.45	16.87
50	6.99	0.37	14.42	18.07
100	10.71	0.60	12.82	19.92
200	26.75	0.85	7.76	22.27
250	56.78	0.93	3.53	29.08
500	96.82	0.96	0.00	29.89
1000	97.88	0.97	0.00	31.12

Bitcoin babble

The acceptance of money as a medium of exchange has been central to our economic evolution. No longer do people need to be self-sufficient, growing their own food, making their own clothes, and building their own furniture. No longer must people barter, trading fruit for fish and chairs for canoes. People can specialize in what they do well and use the money they are paid to buy the things they want.

It hardly matters what is accepted as money—seashells, whale teeth, and woodpecker scalps have all been used. The people of Yap used stones canoed in from hundreds of miles away, even a large stone that had fallen to the bottom of the ocean generations earlier. As long as people are confident that they can use shells, scalps, or stones to buy things, they accept them as payment.

Nowadays, money need not be a physical object, and most of us choose to live in an essentially cashless society. Our income is deposited into our bank accounts and our bills are paid out of our bank accounts. Bills that are not paid electronically are paid by check—not by walking to the water company with a bag full of coins and bills. When we shop for food or gas or go to a restaurant or a movie theatre, we can pay with a debit card or credit card. We only need cash for paying people who won't take checks or credit cards—perhaps buying food at a local farmer's market or paying our children to mow the lawn.

The U.S. has reached the point where cash could be replaced almost completely by a system of electronic debits and credits. Yet, the amount of dollar bills being used continues to grow, particularly large denominations. There are now more than 12.5 billion $100 bills in circulation, up from 2 billion twenty years ago. Divided by the size of the U.S. population, those 12.5 billion $100 bills are 38 Benjamins for every man, woman, and child in the country. Where is all this cash and why do people have it?

Some cash is held as reserves by central banks; some is used by private citizens and businesses outside the United States as a second currency or as a hedge against economic instability. Much of the cash, domestically and abroad, is used for illegal or untaxed activities. Cash transactions are anonymous and can be used to hide clandestine activities from the police and tax authorities, but there are inconveniences.

One of Gary's former students carried around cash for Bulgarian mobsters, with a gun in his pocket. It was scary. Cash can be lost or stolen. Transactions can be videotaped.

Which brings us to bitcoins and other cryptocurrencies. Unlike electronic money that flows through the banking system and is monitored by central banks, bitcoins use cryptography and decentralized control maintained by blockchains. Here is a quotation from the original paper that launched bitcoin:

[W]e proposed a peer-to-peer network using proof-of-work to record a public history of transactions that quickly becomes computationally impractical for an attacker to change if honest nodes control a majority of CPU power.

Doesn't that sound impressively mysterious? Part of the allure of cryptocurrencies is that a great many people worship computers, and very few people understand blockchains. If bitcoin was an accounting system maintained by an army of accountants with pencils and paper, it wouldn't be nearly as cool, or alluring.

As an investment, a bitcoin is no better than a woodpecker scalp or a Yap stone at the bottom of the ocean. Real investments—like stocks, bonds, and apartment buildings—generate real income: bonds pay interest, stocks pay dividends, and apartments pay rent. Bitcoins generate no income whatsoever.

Speculators buy bitcoins because they think they will be able to sell their bitcoins a short while later for an even higher price. This is the Greater Fool Theory—buy at a foolish price and sell to an even bigger fool for a profit. During the Tulip Bulb Bubble, bulbs that might have fetched $20 (in today's dollars) in the summer of 1636 were bought by fools for $160 in January and $2,000 a few weeks later. The prices of exotic bulbs topped $75,000. During the South Seas Bubble in the 1700s, fools bought worthless stock that they hoped to sell to other fools. One company was formed "for carrying on an undertaking of great advantage, but nobody is to know what it is." The shameless promoter sold all the stock in less than five hours and left England, never to return. Another stock offer was for the "nitvender" or selling of nothing. Yet, nitwits bought nitvenders.

Bitcoins are a modern-day nitvender, in that the price of bitcoins is no more related to economic fundamentals than was the price of the South Sea nitvender stock. There is no *there* there. Bitcoin prices are supported by nothing more than the faith that greater fools will pay higher prices.

In 2017, as the Bitcoin bubble picked up speed, the stock price of Long Island Iced Tea Corp. increased by 500 percent after it changed its name to Long Blockchain Corp. At the peak of the bitcoin bubble, a company introduced a cryptocurrency that didn't even pretend to be a viable currency. It was truthfully marketed as a digital token that had "no purpose." Yet, people spent hundreds of millions of dollars buying this nitwit coin.

Figure 2.9 shows the daily price of bitcoins from 2010 through 2018; Figure 2.10 shows the volume of trading. There was a run-up in price and volume in 2017, culminating in nearly simultaneous peaks in December 2017 (a peak price of $19,345 on December 16 and peak volume of 5.2 billion on December 22), followed by sharp declines in price and volume after these twin peaks. These strong co-movements in price and volume are consistent with the view that the run-up and decline were caused by investor speculation.

Suppose, however, that an analyst were to take a deep dive into the data without considering the nature of bitcoins. If the research were done in the summer of 2018, they might look at the period January 1, 2011, to May 31, 2018, a period when bitcoin's price increased from $0.30 to $7530.55, an annual return of 292 percent, and treat bitcoin as a normal investment, like stocks—though much more profitable than stocks, which average only about a 10 percent annual return.

He might find evidence of momentum in bitcoin prices. As Figure 2.9 shows, when the price increased, it tended to keep increasing; when the price fell it tended to keep falling. He might even come up with a trading

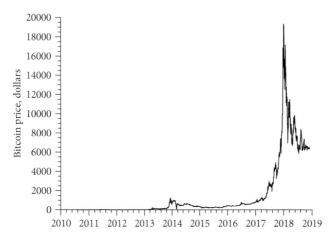

Figure 2.9 The bitcoin bubble

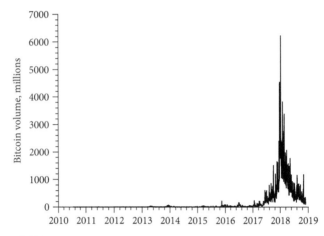

Figure 2.10 Bitcoin excitement

rule based on historical price fluctuations: "buy bitcoin if its price increases more than 20 percent the previous week."

He might also find that bitcoin's price tended to increase after surges in Google searches for the word *bitcoin* and Twitter posts using the word *bitcoin*. He might even find a surge in Google searches for the phrase *bitcoin hack* before the bitcoin price declined, and feel no need to explain why this phrase worked while other phrases he looked at did not work.

Stocks are often valued by looking at the dividend-price ratio—a firm's annual dividend, divided by its price. However, unlike stocks, bitcoins don't generate cash or pay dividends, so the data analyst could create a proxy, like the number of Bitcoin Wallet users, and then calculate a dividend-price ratio by comparing bitcoin prices to the number of Bitcoin Wallet users.

Finally, he might calculate correlations between bitcoin returns and hundreds of other financial variables, and find that bitcoin returns are positively correlated with stock returns in, say, the consumer goods and health care industries and negatively correlated with stocks returns in the fabricated products and metal-mining industries, and think that these discovered correlations are meaningful.

After presenting all these facts, he might argue that bitcoins are an investment well worth owning:

If you as an investor believe that Bitcoin will perform as well as it has historically, then you should hold 6% of your portfolio in Bitcoin. If you believe that it will do half as well, you should hold 4%. In all other circumstances, if you think it will do much worse, then you should still hold 1%.

What would you think? We hope that you are deeply skeptical because you recognize data mining and p-hacking when you see it. This person clearly put data before theory and rummaged through historical data, in the time-honored way of the data clown, and completely ignored the facts that (a) legitimate investments generate cash that rewards investors; and (b) speculative investments that generate no cash are prone to bubbles and crashes.

Applying data mining and p-hacking to stock returns would be bad enough, but it is frankly delusional to treat the 2017 bitcoin price explosion as a normal return that can be analyzed the way one would analyze the returns from bonds, stocks, real estate, and other investments.

This example illustrates why graphs can be illuminating. Bitcoin's extraordinarily high average annual return was not consistent, year after year. If, instead of just calculating the average return, the analyst had drawn a graph like Figure 2.9, he would have seen the extraordinary spike in 2017, followed by the price collapse in 2018. An investor counting on a 200-percent-plus annual return going forward is counting on a run-up like 2017, but much, much larger.

The price momentum and Internet fever documented by the analyst are both exactly what one expects in a speculative bubble, not reliable guides for making money. The number of Bitcoin Wallet users is not a substitute for cash dividends paid to stockholders. This farfetched proxy is reminiscent of the useless metrics (like website visitors) that wishful investors created during the dot-com bubble to justify ever higher stock prices.

The analyst's calculation of hundreds of correlations between bitcoin prices and other variables, and focus on the four most statistically significant correlations, is clearly p-hacking.

We hope that you recognize all the data science pitfalls here, and will be surprised to learn that this study was actually done by a prominent Yale economics professor and a graduate student in the summer of 2018. Even brilliant scientists can stumble into data science pitfalls.

The professor was candid: "We don't give explanations, we just document this behavior." In other words, data before theory.

The 6 percent portfolio recommendation above is based on the assumption that bitcoin's future return will be 200 percent higher than the return on other assets; the 3 percent recommendation is based on a future bitcoin return that is 100 percent higher than on other assets. Their worst possible case ("all other circumstances") is that bitcoin will beat other assets by 30 percent a year.

They do not consider the implications of these assumptions. On May 31, 2018, the last day used in their analysis, bitcoin's price was $7530.55. If, starting at a price of $7530.55, the price were to increase by 100 percent a year, the price would be nearly $8 million ten years hence. With a 200 percent annual increase, the price would be $445 million in ten years' time. These numbers don't pass the straight-face test. We don't know what bitcoin prices will be in 2028, but we are confident that they will be closer to $0 than to $445 million.

They do not even consider the possibility that bitcoin returns might be negative, which is what happens as the air goes out of bubbles. We are writing this on February 12, 2019, with bitcoin's price at $3637.62, down 52 percent since they did their analysis.

In addition to not worshiping math or computers, good data scientists do not worship the advice of professors, just because they are at Ivy League schools. Good data scientists think for themselves and avoid data science pitfalls. They also know that data clowns wearing clever disguises live among us.

After Gary wrote an article warning investors away from bitcoin (at a time when the price was around $5,500), he received an email from a bitcoin enthusiast (Jim) who wanted to bet that the price of bitcoin would be above $1 million in 10 years (November 2028). Gary was tempted, but how could he ensure that Jim would pay up?

Gary came up with a clever proposal:

We agree now to a futures price of $500,000 that you will pay me for one bitcoin ten years from the day the agreement is made. If the market price ten years from now is between $0 and $500,000, I make up to $500,000; if the market price is above $1,000,000, you make $500,000+. From your perspective, expecting the price to be above $1 million, the odds are in your favor. The problem still is: how do I ensure that you will pay me $500,000 ten years from now? So, you buy a 10-year Treasury bond with 2028 payoff of $500,000 and give that bond to a neutral third party who will give it to me in 2028.

Faced with the prospect of paying $500,000 for something that might be worthless, Jim didn't take Gary up on his offer. It's unfortunate, because Gary couldn't lose! He'd buy a bitcoin for $5,500 and, even if he lost the bet and had to sell a bitcoin that was worth $1 million for only $500,000, he would still get a fantastic return on his investment. Good data scientists don't gamble; they invest.

Theory before data

The fundamental problem with data mining is simple:

We think that data patterns are unusual and therefore meaningful.
Patterns are, in fact, inevitable and therefore meaningless.

This is why data mining is not usually knowledge discovery, but noise discovery.

Finding correlations is easy. Good data scientists are not seduced by data-mined patterns because they don't put data before theory. They do not commit Texas Sharpshooter Fallacies or fall into the Feynman Trap.

The Second Pitfall of Data Science is:

Putting Data Before Theory

Worshiping Math

> "We made an orthogonal linear transformation
> of the data and identified the loading vector with the
> largest eigenvalues for a principal components regression."
> "Great!"

The marketing VP had absolutely no idea what the data scientist said, but it sounded smart. When an expert says something that is confusing, but sounds impressive, nodding politely is sometimes less embarrassing than admitting ignorance.

The Dr. Fox effect

In 1970 an actor, Michael Fox (not to be confused with Michael J. Fox) was given the identity *Dr. Myron L. Fox* for a hoax at the University of Southern California. Dr. Fox gave lectures to a group of psychiatrists and psychologists with MDs and PhDs on the topic "Mathematical Game Theory as Applied to Physician Education," a bogus subject chosen to eliminate the possibility that the audience might know something about the topic.

Deborah Merritt, a law professor, summarized the "Dr. Fox effect":

The experimenters created a meaningless lecture on "Mathematical Game Theory as Applied to Physician Education," and coached the actor to deliver it "with an excessive use of double talk, neologisms, non sequiturs, and contradictory statements." At the same time, the researchers encouraged the actor to adopt a lively demeanor, convey warmth toward his audience, and intersperse his nonsensical

The 9 Pitfalls of Data Science. Gary Smith and Jay Cordes. Oxford University Press (2019). © Gary Smith and Jay Cordes 2019. DOI: 10.1093/oso/9780198844396.001.0001

comments with humor …. The actor fooled not just one, but three separate audiences of professional and graduate students. Despite the emptiness of his lecture, fifty-five psychiatrists, psychologists, educators, graduate students, and other professionals produced evaluations of Dr. Fox that were overwhelmingly positive.

If we don't understand something, it is tempting to think that the speaker or author must be smarter than us. So it is with math.

Math rules

Mathematicians love math and many non-mathematicians are intimidated by math. This is a lethal combination that can lead to the creation of wildly unrealistic mathematical models.

We were both math majors in college. Gary has a PhD in Economics from Yale and teaches finance and statistics; Jay has a Master's degree in Data Science from UC Berkeley. We use math virtually every day of our lives. We love math, but we also know that the seductive allure of math can lead people to build mathematical models that are intrinsically gratifying, but pointless. As Warren Buffett once warned, "Beware of geeks bearing formulas."

When Gary was an assistant professor at Yale, he attended a lecture by a distinguished mathematical economist who began his talk by declaring, "Making whatever assumptions are needed…." A Nobel Laureate in the audience got up and left the room.

The guest speaker had it backwards. A good mathematical model starts with plausible assumptions and then uses mathematics to derive the implications. A bad model focuses on the math and makes whatever assumptions are needed to facilitate the math.

Gary attended another lecture in which the speaker started by specifying two assumptions that he needed to prove a theorem about economic behavior. As the speaker carefully worked through a complicated series of theorems, Gary was sitting next to a Nobel Laureate who was furiously scribbling on a notepad. Gary thought that he was checking the math. When the speaker finished his presentation and asked for questions from the audience, the Nobel Laureate said, "I want to talk about your second assumption."

That is exactly right. The plausibility of the assumptions is more important than the accuracy of the math. There is a well-known saying about data analysis: "Garbage in, garbage out." No matter how impeccable the statistical analysis, bad data will yield useless output. The same is true of mathematical models that are used to make predictions. If the assumptions are wrong, the predictions are worthless.

Good data scientists know that they need to get the assumptions right. It is not enough to have fancy math. Clever math with preposterous premises can be disastrous.

Fat tails

Figure 3.1 shows that the probabilities for the number of heads when 1000 coins are flipped has a familiar bell-shaped normal distribution.

Many propositions in finance are based on the mathematically convenient assumption that, like coin flips, stock returns are normally distributed. However, that assumption is more convenient than credible. Stock returns are not like coin flips, with each outcome equally likely to be a head or a tail, no matter how previous flips landed. Investors often overreact to good and bad news and they are prone to chasing trends, so that stock returns are sometimes more extreme—good or bad—than would be true of a normal distribution.

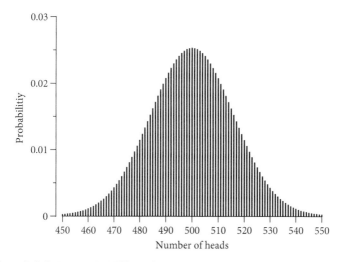

Figure 3.1 Probabilities for 1,000 coin flips

Table 3.1 *Actual number of big days and expected number with normal distribution*

Daily return	Expected number	Actual number
Greater than 10%	0.000011	439
Greater than 5%	270.1	3810
Lower than –5%	270.1	3021
Lower than –10%	0.000011	337

To demonstrate this, we looked at the daily returns on each stock in the Dow Jones Industrial Average from October 1, 1928, when the Dow was expanded from 20 to 30 stocks, through December 31, 2015, a total of 22,965 trading days. Table 3.1 compares the actual number of big days (when a stock went up or down more than 5 or 10 percent) with the expected number of big days if stock returns were normally distributed. There were many more big days than expected. These are called *fat tails*, because the outer edges (the tails) of the distribution of stock returns are fatter than the tails of the normal distribution.

The stock market as a whole also has fat tails. On October 19, 1987, the S&P 500 index fell 20.3 percent, which, if the data were normally distributed, should not happen once in a trillion years. And two days later, on October 21, 1987, the market went up 9.7 percent, which also shouldn't happen in a trillion years. Something that shouldn't happen once in a trillion years happened twice in three days.

Something similar happened twenty years later, in 2007. The head of quantitative equity strategies for Lehman Brothers lamented that, "Events that models only predicted would happen once in 10,000 years happened every day for three days."

This is also called the *black swan* problem. The English used to believe that black swans did not exist because all the swans they had ever seen or read about were white. However, no matter how many white swans we see, these sightings can never prove that all swans are white. Sure enough, a Dutch explorer found black swans in Australia in 1697.

The fact that, up until October 19, 1987, the S&P 500 had never risen or fallen by more than 9 percent in a single day, did not prove that it could not do so. Even worse, how can we make a credible estimate of the chances of an event so rare that it has not yet occurred? The convenient assumption that stock returns are normally distributed assumes the problem away, an assumption that blew up many risk-management models that woefully underestimated the chances of huge swings in stock prices.

The presumption that stock returns are normally distributed has also distorted many stock valuation models. For example, in the 1960s, two finance professors, Fischer Black and Myron Scholes, developed what has come to be known as the Black–Scholes model for valuing options. A stock call option gives the purchaser the right to buy a stock at a specified price up until a specified expiration date. A put option gives the owner the right to sell at a specified price.

Black and Scholes used some advanced mathematics to derive the "correct" prices of call options. Scholes received a Nobel Prize in Economics; Black was ineligible because he was deceased.

Unfortunately, almost every assumption used to derive the Black–Scholes model is wrong! Remember the professor who began his talk by blithely saying, "Making whatever assumptions are needed." That is exactly what Black and Scholes did. They made several completely unrealistic assumptions so that their math would work. For example, they assumed that stocks returns are normally distributed. They aren't. They have fat tails.

The model is elegant mathematics, but of little practical use and potentially very misleading. Espen Haug and Nassim Taleb, two experienced options traders, say that they used the Black–Scholes model for a while, but soon figured out its fatal flaws. They and other options traders have stopped using it.

The Black–Scholes model is hardly an exception. Many, many models used by data scientists are based on the normal distribution, which is sometimes a useful approximation to reality and, other times, a perilous distortion of reality. Good data scientists think about what they are modeling before making assumptions.

Predicting the Super Bowl

A sportswriter created the Super Bowl indicator, using the outcome of the U.S. football championship game to predict whether the stock market would go up or down that year. He wanted to demonstrate in a humorous, but persuasive, way that statistical patterns can be meaningless. To his surprise, many people took the Super Bowl indicator seriously, and still do, even though it hasn't done better than a coin flip in recent years.

Let's go the other way, using nonsense data to predict the Super Bowl score, the AFC team minus the NFC team. We will use data for five Super Bowls, 2009–2013, to estimate our model, and data for the next five Super Bowls, 2014–2018, to validate the model. We won't sift through hundreds of potential variables or make up random variables. Promise! We already know that data mining hundreds of variables—made up or real—will yield a good model that is great in-sample and worthless out-of-sample. Our point here is different.

We put our heads together and came up with these four variables, all recorded the year before the Super Bowl:

High temperature in London on June 30
Average price of tea worldwide on June 30
Average number of letters in the names of Physics Nobel Prize winners
Number of points earned by the soccer team winning the Premier League.

The outcome of outdoor sporting events can be affected by the weather, so we picked the temperature in London on June 30, which surely has nothing to do with the outcome of a football game played in the United States many months later. We remembered the proverbial "price of tea in China," but didn't think we could get those data, so we settled on the average price of tea internationally. Nobel Laureates in physics seem pretty far removed from professional football, especially the length of their names, so that appealed to us. Finally, we wanted to use English football to predict American football, so we used the number of points earned by the Premier League winner.

Bear with us. There is method in our apparent madness.

We fit the model to the 2009–2013 data and Figure 3.2 shows our model's success: the line showing the actual score difference is right on top of the line showing the predictions. Super Bowl scores went up and down, but our model tracked them perfectly. There is a 1.0 correlation between the actual score difference and the predictions made by our nonsense model.

Our dirty little secret

Suppose that we want to predict the score difference for only two Super Bowls (2009 and 2010) and we use one variable, London weather. Figure 3.3 shows that there is a perfect correlation. The June 30 high temperature in London predicts the Super Bowl score months later exactly.

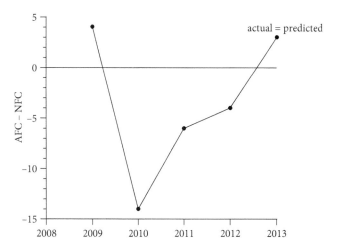

Figure 3.2 Using nonsense to predict Super Bowl scores, 2009–2013

The trick is that there is always a perfect linear relationship between any two points in a scatterplot. We could have chosen any of our nonsense variables, or any other variable. There will be a perfect linear relationship, because any two points lie on the straight line connecting them. That is just a mathematical fact and tells us nothing about whether the two variables are systematically related to each other.

We used two data points on a two-dimensional graph so that we see the folly that applies to more complicated models using more data. Figure 3.3 uses one variable to give a perfect fit to two observations. Two variables give a perfect fit to three observations; four variables give a perfect fit to five observations.

That's why we chose four variables to predict the margin of victory in five Super Bowls, and were confident they would fit the data perfectly even if they were nonsense. This is an extreme example of what is called *overfitting* the data. In any empirical model, we can improve the model's explanatory power by adding more and more variables—in extreme cases, to the point where the fit is perfect. It hardly matters whether the variables make sense or not.

This is also known as the *kitchen-sink* approach to modeling: throw every variable but the kitchen sink at the model. The inescapable problem is that even though the model may fit the original data very well, it is

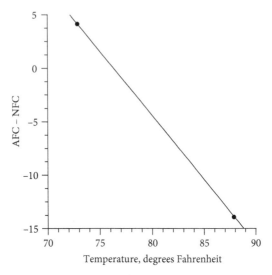

Figure 3.3 London temperatures and the Super Bowl

useless for predictions using new data. Our four nonsense variables won't help us predict scores in other Super Bowls.

Figure 3.4 shows that our model flopped spectacularly out-of-sample, predicting that the AFC would win by 46 points in 2015 (they won by 4) and lose by 64 points in 2016 (they won by 14). The out-of-sample correlation between the model's predictions and the actual margin of victory was −0.25.

Over and over again, we have spoken with intelligent and well-intentioned people who do not fully appreciate how easy it is to find coincidental patterns and relationships, and how a kitchen-sink approach exacerbates the problem. Many are vaguely aware of the possibility of spurious correlations, but nonetheless believe that the mere existence of statistical patterns and relationships is sufficient proof that they are real.

In 2017 Greg Ip, the chief economics commentator for the *Wall Street Journal*, interviewed the co-founder of a company that develops AI applications for businesses. Ip paraphrases the co-founder's argument:

If you took statistics in college, you learned how to use inputs to predict an output, such as predicting mortality based on body mass, cholesterol and smoking. You added or removed inputs to improve the "fit" of the model.

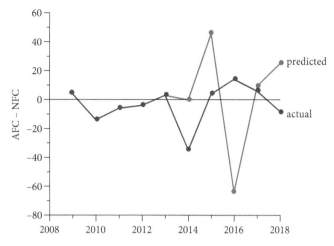

Figure 3.4 Using nonsense to predict Super Bowl scores out-of-sample

No! What statistics students *should* learn in college is that it is perilous to add and remove inputs simply to improve the fit. Rummaging through data looking for the best fit is mindless data mining—and the more inputs considered, the more likely it is that the selected variables will be spurious.

Statistical correlations are a poor substitute for expertise. The best way to build models of the real world is to start with theories that are appealing and then test these models. Models that make sense can be used to make useful predictions.

Nonlinear models

In addition to overfitting the data by sifting through a kitchen sink of variables, data scientists can overfit the data by trying a wide variety of nonlinear models.

For example, the theory that household spending is related to income is compelling. Households with more income generally tend to spend more, though the relationship is not perfect because other factors matter too, like the number of people in the household, the ages of the household members, and household wealth. Figure 3.5 shows a simple scatterplot

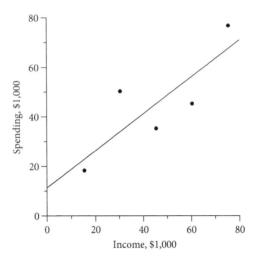

Figure 3.5 A reasonable linear model

using hypothetical data on household income and spending. It seems perfectly reasonable to fit a straight line to these data and to interpret the variation in the points about the line as the effects of factors that are not in the model. The straight line should make reasonably accurate predictions, at least for the range of incomes considered here.

Figure 3.6 shows what happens when a data-mining algorithm overfits the data by using a nonlinear model to fit the data perfectly. Despite the perfect fit, the nonlinear model's predictions for other values of income are surely wildly inaccurate and sometimes bizarre. According to this nonlinear model, households with $30,000 income spend more than do households with $50,000 or $60,000 income, and when income goes up, spending sometimes goes up and sometimes goes down.

When the data can be plotted in a simple scatter diagram, like Figure 3.6, it is obvious that, despite the great fit, the nonlinear model is worse than a simple linear model. Problems arise, however, when a model has many variables, and we can't use a simple scatter diagram like Figure 3.6 to reveal the model's flaws. What is certain is that the success of a model cannot be assessed solely by looking at how closely the model fits the data,

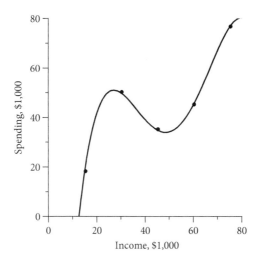

Figure 3.6 An implausible nonlinear model

Living with uncertainty

There were 32 teams at the 2018 World Cup finals in Russia—the host country plus 31 teams that advanced through qualifying rounds. Shortly before the matches began, a team of German and Belgian researchers announced that they had used a sophisticated computer algorithm to predict the winner. The algorithm said that Spain had a 17.8 percent chance of winning, Germany a 17.1 percent chance, and Brazil a 12.3 percent chance. Altogether, there was a 47.2 percent probability that one of these three countries would be champion.

A news report concluded that, "The results of this study demonstrates that machine learning models can be used to predict the outcome of major sporting events." Well, yes, the model made predictions. The hard part is making reliable predictions. Germany did not make it to the final 16, Spain did not make it to the final 8, and Brazil did not make it to the final 4.

Top-level soccer matches are notoriously capricious and are often decided by a single goal. Sometimes the ball bounces off a goal post and into the goal; others times it bounces wide. Sometimes a referee awards a penalty; other times, the referee says, "Play on."

When uncertainty is the order of the day, models that purport to be incredibly accurate are not credible. There may have been extensive data mining or the model may overfit the data. Occasionally, the reported results are simply untrue.

Gary recently refereed a paper that claimed a 0.99 correlation between daily stock returns and the model's predicted stock returns. Gary has been investing and teaching investing for decades, and he knew that this claim was not credible. As Keynes observed, the stock market is buffeted by fear, greed, and other inexplicable "animal spirits." The 0.99 correlation was not believable. If the model was really that accurate, the authors would have been busy becoming billionaires instead of writing academic papers. Gary looked through the paper and there was no clear description of the methodology. The kindest interpretation is that the authors obtained an unrealistic in-sample correlation by overfitting, and that the out-of-sample predictions would be garbage.

Good data scientists know that some predictions are inherently difficult and we should not expect anything close to 100 percent accuracy. It is better to construct a reasonable model and acknowledge its uncertainty than to expect the impossible.

TradeALot

A data-driven hedge fund (TradeALot) asked a data scientist ("Rick") to develop a stock-trading system based on historical stock prices. Rick wrote a program that used dozens of variables to create buy or sell signals and was "tuned" by iterating through thousands of possible coefficients for each of the variables. Rick's program fit models to the historical data the same way we fit our nonsense Super Bowl model to the data.

Nothing in Rick's model had anything to do with the specific stock being predicted. The model could have just as easily been tuned to fit coin flips and discover what looked like a winning trading strategy.

Rick's model discovered incredibly profitable trading strategies, like the one shown in Figure 3.7, because, with dozens of variables, there were always apparent ways to profit from patterns in historical data. The data mining was out of control, yet the CEO shouted, "We're going to be rich!"

The CEO offered to pay Rick a percentage of the trading profits when the system was used. Rick was skeptical about the system actually working, so his counter-offer was $100 an hour: "If you think it's going to make

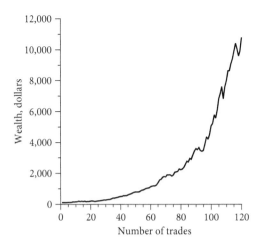

Figure 3.7 Easy money!

a ton of money, then you should be ecstatic to pay me only $100 an hour and pocket the rest." The CEO didn't bite, so Rick found a different job, suspecting that, deep down, the CEO knew the strategy was garbage.

There is an enduring saying in finance: "If it sounds too good to be true, it probably is." The model in Figure 3.7 would have turned $100 into $10,792 after 120 trades in one day. At that rate, wealth would have gone from $100 to over $1 trillion in one week. Talk about too good to be true!

Not surprisingly, the TradeALot trading systems that seemed to work so well with historical data usually stunk when they were used with live data. Convenient excuses were readily available. There was "slippage" with out-of-sample data. The market had changed and the opportunity had passed. The market was nuts. They never considered the more likely explanation that models created by mining data and overfitting data are inherently flawed.

Data-mined stock market systems come and go as models are fit to historical data and tested on live data. Just by chance, some systems may be lucky with live data for a while. Sometimes, after a coin lands heads five times in a row, the streak continues and it comes up heads a sixth time and a seventh time. It is still just a random coin flip—and the chances of coming up heads on the next flip is still 50–50.

TradeALot's clients would have been better off paying monkeys peanuts to flip coins, because TradeALot charged a lot for its flawed models: 2

percent of assets, no matter how the fund did, and 20 percent of profits when there happened to be profits. TradeALot didn't lose money when its clients lost money; they just lost the clients.

Nice work if you can get it. Essentially, money for nothing.

Statistical overconfidence

Suppose you want to compare four web page layouts to see which layout is most likely to get visitors to click on ads. You use software that randomly sends visitors to one of the four layouts. Then you track the clicks and visits, and calculate the conversion rate (clicks per visitor). The page with the highest conversion rate is the winner.

You want to make sure that the observed differences are statistically significant, so you calculate p-values, the probability that the observed differences would be this large by chance alone. For example, if someone claimed that she could predict the outcomes of coin flips and got 55 of 100 correct, the probability of doing that well by random guessing is 0.18; so her p-value is 0.18.

Typically, an event is considered statistically significant if the p-value is less than 0.05. A coin predictor would have to get at least 59 correct to demonstrate statistically significant evidence of her prediction prowess. This seems like a reasonable hurdle; however, about 1 out of 20 guessers will get 59 or more correct just by luck alone, and mislead us into concluding that they can predict coin flips. That's why it's a Texas Sharpshooter Fallacy to do lots and lots of tests and only report the statistically significant ones.

Jay discovered that several unrelated companies used a statistical approach that caused them to draw incorrect conclusions far more often than 5 percent of the time. They assumed that the p-values for conversion rates could be calculated the same way as with coin flips, by using something called the binomial distribution. The calculations are complicated because they compared two layouts, which is analogous to comparing two coin-flip predictors to each other, rather than a single person to a 50 percent guess probability, but they used the correct binomial formula for their calculations. Their software plugged their conversion data into the binomial formula and chugged out p-values.

Jay noticed some peculiarities in their results. For example, the two layouts in Figure 3.8 gave the same p-values in comparison to a third layout.

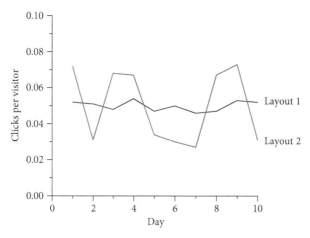

Figure 3.8 Two equally convincing layouts?

Jay knew intuitively that the data for Layout 2 were much more volatile and should not be as persuasive as the data for Layout 1.

Figure 3.9 shows another example of two layouts that gave the same p-values in comparison to a third layout. Layout 2 was consistently inferior, then had a spike on Day 8. How could the data for these two layouts be equally persuasive when Layout 2's success hinged solely on one fluky day?

All the layouts in Figures 3.8 and 3.9 have the same number of visitors and the same average conversion rate, like two coin-flip predictors who flip the same number of coins and get the same number correct, so the binomial distribution considers the results to be identical. Yet, the results are clearly not identical. Jay realized that the math was right, but the assumptions were wrong. Like the mathematical economist who began his lecture by declaring, "Making whatever assumptions are needed," these companies didn't think about whether their assumptions were correct.

The binomial distribution applies to things like coin flips, where every flip has the same constant probability of occurring. Jay saw several problems. First, click rates are not necessarily constant (like coin-flip probabilities), but can vary substantially day to day. The daily volatility shown in Figures 3.8 and 3.9 should be taken into account, instead of ignored.

Second, the binomial distribution assumes that the outcomes are independent, the way that a coin flip doesn't depend on previous flips.

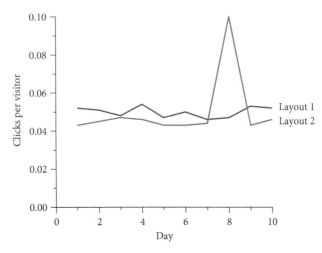

Figure 3.9 Are outliers all that matter?

However, there are several ads on most web pages, and a comparison shopper or a "bot" (a computer program) crawling the Internet could visit a page once, and click on multiple ads. Clicks are not independent for a comparison shopper or a bot. The binomial distribution assumes, for example, that 10 different people went to a page and clicked on an ad when, in fact, one person (or one bot) may have gone to the page and clicked on 10 ads. That's how the spike on Day 8 in Figure 3.9 happened. The luck of the draw sent some comparison shoppers or bots to the page with Layout 2, which surely should not be interpreted as decisive evidence favoring Layout 2.

There's another problem. The conversion rate is the number of clicks per visitor. Since there are several ads on each web page, there can be more clicks than visitors, causing the conversion rate to go above 1. This causes the binomial calculation to blow up because you can't have 12 heads in 10 coin flips.

A telling sign that your mathematics don't match reality is when one of your equations crashes. The companies doing these calculations used a hack to deal with their blown-up binomial distributions: they capped the conversion rate at 1. This is like saying that if a model reports 12 heads in 10 flips, we will pretend that there were only 10 heads—instead of thinking about why the model reported 12 heads.

It evidently didn't bother them that a hack was necessary—just make it work! If they had thought about why their calculations were blowing up, they might have realized that the binomial distribution was not appropriate, and figured out how to calculate meaningful p-values.

In addition, their p-value calculations assume that they are only doing one test. They were making a Texas Sharpshooter Fallacy when they made multiple tests. With four competing layouts, there is a 0.26 probability that chance alone will give at least one p-value less than 0.05. With six competing layouts, the probability is 0.64. With eight layouts, it is essentially certain.

There was another, more subtle Texas Sharpshooter Fallacy. These companies ran their tests for several days, calculating p-values every day, and declared a winner when a p-value dropped below 0.05. They were making multiple tests, not only across layouts but across days, which greatly increased their chances of a misleadingly low p-value. No wonder the layouts they selected usually turned out to be disappointing when they were rolled out.

Figure 3.10 shows an example of one test where blue seemed to be the clear winner with a p-value that dipped below 0.05 after 707 visitors. However, Figure 3.11 shows that when they let the test run a little longer,

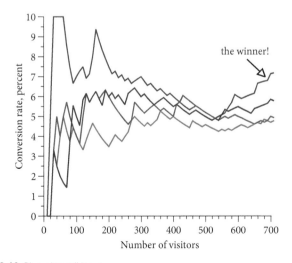

Figure 3.10 Blue wins; roll it out

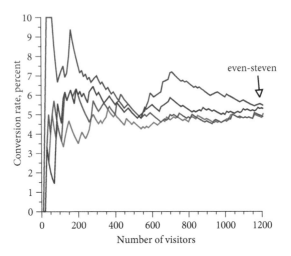

Figure 3.11 Even-steven

blue's edge disappeared. If we p-hack across days, as well as across colors, it increases the chances that, by luck alone, a meaningless pattern will temporarily appear to be significant.

The binomial distribution is elegant mathematics, but it should be used when its assumptions are true, not because the math is elegant. Good data scientists think about the assumptions before they figure out the math.

Principal Components Regression

Look up *principal components regression* on the Internet and you will find lots of equations that make mathematicians smile and give others headaches. It is a nice example of a data science tool that is simultaneously impressive and intimidating, but ultimately problematic.

Data scientists are often asked to build predictive models based on a long list of potential explanatory variables. Principal components regression (PCR) replaces a large number of variables with a small number of weighted averages of these variables. The problem is that the weights depend on the correlations among the candidate variables and have *nothing at all* to do with the variable being predicted. Irrelevant variables may be given larger weights than truly important variables.

To demonstrate this, we created some hypothetical data for predicting the dollar value of accident claims. Two of the variables, the number of traffic tickets and the number of miles driven each year, really do matter. The other two, the person's phone number and the day of the year the person was born, do not matter at all.

We used half the data to estimate the model and the other half to test the model. The first thing PCR did was construct two weighted averages, called the principal components, of the four explanatory variables, with the weights determined by the correlations among the four variables, with no consideration whatsoever for how they might be related to accidents!

Next, PCR estimated a model that fit the accident data as well as possible using the two principal components as explanatory variables. The end result turned out to be a model in which the two variables that don't matter at all were the most important, and the number of miles driven was predicted to have a negative effect on accident claims, when, in fact, it has a positive effect.

The PCR model was then used to make out-of-sample predictions for the data that had been set aside for this purpose. Table 3.2 shows that the out-of-sample prediction errors were much larger than the in-sample errors, no doubt because the model's estimated coefficients were so inaccurate.

For comparison, a naive model that completely ignored the explanatory variables and simply predicted that everyone will have the same accident claims had an average out-of-sample prediction error of 40.56. The PCR model was slightly worse than useless for making predictions.

This a simple example of how a mathematically elegant procedure can generate worthless predictions. Principal components regression is just the tip of the mathematical iceberg that can sink models used by well-intentioned data scientists. Good data scientists think about their tools before they use them.

Table 3.2 *Principal components' average prediction errors*

In-sample	Out-of-sample
12.34	42.93

Blinded by math

Data-mining tools, in general, tend to be mathematically sophisticated, yet often make implausible assumptions. Too often, the assumptions are hidden in the math and the people who use the tools are more impressed by the math than curious about the assumptions.

Instead of being blinded by math, good data scientists use assumptions and models that make sense. Good data scientists use math, but do not worship it. They know that math is an invaluable tool, but it is not a substitute for common sense, wisdom, or expertise.

The Third Pitfall of Data Science is:

Worshiping Math

Worshiping Computers

"Ask the computer if I should buy GE stock."
"I've never owned a share of stock in my life."
"I'm not asking you. I'm asking the computer. What does the computer think of GE?"

When Gary first started teaching economics in 1971, his wife's grandfather ("Popsie") knew that Gary's PhD thesis used Yale's big computer to estimate an extremely complicated economic model. Popsie had bought and sold stocks successfully for decades. He even had his own desk at his broker's office where he could trade gossip and stocks.

Nonetheless, he wanted advice from a 21-year-old kid who had no money and had never bought a single share of stock in his life—Gary—because Gary worked with computers. "Ask the computer what it thinks of GE."

This naive belief that computers can think has been around ever since the first computer was invented more than 100 years ago by Charles Babbage. In his autobiography, he recounted that, "On two occasions I have been asked [by members of parliament], 'Pray, Mr. Babbage, if you put into the machine wrong figures, will the right answers come out?'" Babbage lamented that, "I am not able rightly to apprehend the kind of confusion of ideas that could provoke such a question."

Even today, when computers are commonplace, many well-meaning people still cling to the misperception that computers can think, indeed,

The 9 Pitfalls of Data Science. Gary Smith and Jay Cordes. Oxford University Press (2019).
© Gary Smith and Jay Cordes 2019. DOI: 10.1093/oso/9780198844396.001.0001

that computers are smarter than us. The reality is that computers still can't think like us, though they do seem to have gotten into our heads.

It is true that computers know more facts than we do. They have better memories, make calculations faster, and do not get tired like we do. Robots far surpass humans at repetitive, monotonous tasks like tightening bolts, planting seeds, searching legal documents, and accepting bank deposits and dispensing cash. Computers can recognize objects, draw pictures, and drive cars. You can surely think of a dozen other impressive—even superhuman—computer feats.

It is tempting to think that because computers can do some things extremely well, they must be highly intelligent, but being useful for specific tasks is very different from having a general intelligence that applies the lessons learned and the skills required for one task to more complex tasks, or completely different tasks.

Our awe of computers is not a harmless obsession. If we think computers are smart, we will be tempted to let them do our thinking for us—with potentially disastrous consequences. Computers cannot distinguish good data from rubbish, genuine from spurious, or sensible from nonsense. Computers do not think in any meaningful sense of the word.

Anthropomorphization

Humans often anthropomorphize by assuming that animals, trees, trains, and other non-human objects have human traits. We tell stories in which pigs build houses that wolves blow down, foxes talk to gingerbread men who have run away from home, and bears are upset that a girl sits in their chairs, eats their porridge, and sleeps in their beds.

These are enduring stories because we are so willing, indeed eager, to assume that animals (and even cookies) have human emotions, ideas, and motives. In the same way, we assume that computers have emotions, ideas, and motives. They don't.

Many people are fascinated and terrified by apocalyptic science-fiction scenarios in which computers decide they must eliminate the one thing that might disable them: humans. The success of Jay's favorite movies, *The Terminator* and *The Matrix*, has convinced many that this is our future and it will be here soon. Even luminaries such as Elon Musk have warned of robotic rebellions. In 2014, Musk told a group of MIT students, "With artificial intelligence we are summoning the demon.... In all those stories where there's the guy with the pentagram and the holy water, it's like yeah he's sure he can control the demon. Didn't work out." Three years later, Musk posted a photo of a poster with the ominous warning, "In the end, the machines will win."

The idea of an imminent computer takeover is pure fantasy. Computers do not know what the world is, what humans are, or what survival means, let alone how to survive. The real danger is that, because we think computers are smarter than us, we will trust them to make important— even life and death—decisions. Which mortgage applications should be approved? Which job applicants should be hired? Which people should be sent to prison? Which medicine should be taken? Which targets should be bombed?

Black boxes

To make matters worse, computer algorithms are often hidden inside black boxes that make the models inscrutable: inputs are fed into the algorithm, which provides output without human users having any idea how the output was determined. No one knows why a black-box algorithm concluded that this stock should be purchased, why this job applicant should be rejected, why this patient should be given this medication, why this prisoner should be denied parole, or why this building should be bombed.

For a black-box stock trading algorithm, the inputs might be data on stock prices, the number of shares traded, interest rates, the unemployment rate, the number of times the word "excited" is used in tweets, sales of yellow paint, and dozens of other variables. The output might be a decision to buy or sell 100 shares of GE stock. How did the computer make that decision? Who knows? Who cares? Certainly not Popsie. We all should care.

Users who are willing to let black-box trading algorithms buy and sell stocks for them do not know the reason for these decisions, but they are untroubled because they trust the black box. They think that their computers are smarter than they are. This is meant to be reassuring, but it is frightening.

The computer poker competition

In 2007, the world's best computer poker programs competed heads-up at a no-limit Texas Hold 'Em competition at the Association for the Advancement of Artificial Intelligence in Vancouver, Canada. Anyone could enter, but it was assumed that the competition would be won by a team of researchers from a top university, such as the University of Alberta or Carnegie Mellon, using brilliant coders to write state-of-the-art artificial intelligence software. One of the Alberta researchers had written his doctoral thesis on, "Algorithms and Assessment in Computer Poker." The Carnegie Mellon team boasted that they used "automated abstraction equilibrium computation techniques." The chances that a hobbyist could create a computer program that would be competitive with these teams was slim, at best.

A hobbyist had shown promise the prior year, finishing second only to the University of Alberta in the limit Heads-Up poker competition. This guy was a friend of a friend of Jay's and came to a poker night at Jay's house in 2006. Jay urged him to rewrite his bot for the no-limit rules of the 2007 competition and offered to serve as a sparring partner.

The program was small, less than 1 MB in size. Carnegie Mellon's program was 8 GB, nearly 10,000 times bigger. And yet, the tiny algorithm went undefeated against its nine opponents, blindsiding the competition. Michael Bowling, the leader of the University of Alberta's Computing Science Department, said, "They are going up against top-notch universities that are doing cutting-edge research in this area, so it was very impressive that they were not only competitive, but they won."

Data scientists can be lured into trusting the latest high-powered algorithm simply because it is new and high powered when a simple, carefully crafted approach might work better. The research teams in the poker competition may be excused for focusing on complicated math (they had to produce research papers after all), but data scientists should always start with the simplest approach that makes sense and only ramp up the complexity when it's necessary.

We are not anti-computer. We use computers every day of our lives. Jay was a software developer for 11 years, and Gary has written nearly 100 research papers, almost all using computers for mathematical calculations, statistical analyses, and computer simulations. Many of the calculations could not have been done in one lifetime without computers. But there is a difference between using computers to do amazing things and trusting computers simply because they are capable of doing amazing things. There is a fundamental difference between mindless calculations and critical thinking.

Monte Carlo simulations

Bobby Layne, a Hall of Fame football quarterback, once said, "My idea of a full life is to run out of money and breath at the same time." This pithy quip dramatizes the reality that the elderly want to enjoy their retirement years, but fear that they may exhaust their savings prematurely. Whether or not they will outlive their wealth depends on two crucial decisions: how much they spend each year and the investments they make.

Gary worked with an investment advisor to analyze some of the financial risks faced by retirees. The two main client concerns were the bequests they would leave to their heirs and the probability that the inflation-adjusted value of their wealth would, at some point, fall far below its initial value—which would not only threaten their bequest but could possibly cause them to run out of money before they run out of breath.

Many financial advisors use an extremely simplistic retirement planning model that assumes that people know the day they will die and also know the return on their investments. The advisors then use a straightforward mathematical formula to determine how annual spending affects their bequest.

Two obvious flaws are that people do not know how long they will live or how well their investments will do. These uncertainties are precisely why people are concerned about outliving their wealth. These simple retirement models assume the problem away!

Why did these models use such obviously unrealistic assumptions? Because, without these assumptions, there is no straightforward mathematical formula for making the necessary calculations.

With powerful computers, there is an attractive alternative, called Monte Carlo simulations (named after the famous gambling mecca). Monte Carlo simulations handle uncertainty by using a computer's random number generator to determine outcomes. Done over and over again, the simulations show the distribution of the possible outcomes.

Here, the users of Gary's model specify whether there is a single person or a couple, their age(s), their initial wealth, their income, their spending, and what percent decline in wealth they consider to be a substantial shortfall (quarterback Bobby Layne would say 100 percent). The model uses government data on the projected death rates (for example, the probability that a female who is 82 in 2026 will die that year), and the program allows these death rates to be adjusted if the users think they are unusually healthy or unhealthy.

In each year of the simulation, the program uses the mortality probability and a random number generator to determine whether the person dies that year. If there is a couple, separate calculations are made for each person and, if one person dies, spending is assumed to drop by a specified percentage, say 25 percent.

Retirement planning models typically consider two assets—corporate stock and U.S. Treasury bonds—and Gary's model does too, with the percentage of wealth invested in stocks varying from 0 to 100, and the computer's random number generator determining the returns each year on the bonds and stocks in the portfolio.

Finally, users specify how much they will spend each year as a fraction of their initial wealth (the "withdrawal rate"), with spending adjusted each year for inflation. For whatever assumptions are made by users, the model makes one million simulations.

Figure 4.1 shows the results for the case of a couple, each age 65, with $1 million in assets and no outside income, who consider a 50 percent drop in wealth to be a shortfall.

The horizontal axis shows the shortfall probability—the fraction of the simulations in which inflation-adjusted wealth fell below $500,000 at some point. The vertical axis shows the median bequest. In half the simulations, the bequest was larger than this; in half, the bequest was smaller. The two lines are for the cases of 3 percent and 5 percent withdrawal rates. With $1 million in assets, these correspond to spending $30,000 or $50,000 annually. Each line ranges from a portfolio with 0 percent stocks to one with 100 percent stocks.

The beauty of these Monte Carlo simulations is that they allow users to see the probabilistic consequences of their decisions, so that they can make informed choices.

How could we do any of this without Monte Carlo simulations? We couldn't. Monte Carlo simulations are one of the most valuable applications

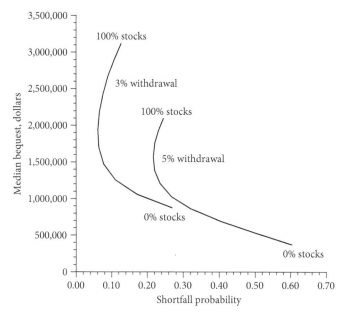

Figure 4.1 The tradeoff between bequest and shortfall probability

of data science because they can be used to analyze virtually any uncertain situation where we are able to specify the nature of the uncertainty: medical care, sporting events, political campaigns, investing, business decisions, and much more.

Just checking

Even when a mathematical analysis is possible, Monte Carlo simulations can be used check the math. For example, at the start of each semester, Gary asks the students in his statistics classes to guess the outcomes of 10 coin flips. He then asks the students to calculate the longest streak in their guesses; for example, 3 consecutive heads or 4 consecutive tails. There is roughly a 47 percent chance of a streak of 4 or longer, so, if the students were randomly guessing, almost half should have streaks this long. In practice, very few do, evidently because they mistakenly believe that heads and tails must even out, making streaks unlikely.

Gary remembered this 47 percent probability from having done this experiment for decades, but he couldn't recall the source of that number, and he wondered if he had misremembered it. He started working out the math and it was intimidating, so he wrote a Monte Carlo program to simulate flipping 10 coins a billion times, and found that the correct percent is 46.48. A simple computer program was able to solve a difficult mathematical problem and confirm that nearly half the time 10 coin flips will have a streak of 4 or more consecutive heads or tails.

Jay had a more complicated example in which direct calculations validated a simulation, and vice versa. Along with Barry A. Balof, Joseph Kisenwether, and the "Mathemagician" Arthur T. Benjamin, he coauthored a mathematical paper on a paradox involving the game of Bingo.

A Bingo card (see Figure 4.2) has 24 randomly selected numbers out of 75 possibilities, 1 to 75, and a free space in the center of the card. A caller randomly selects numbers that the players mark on their cards until someone has five numbers in a row—horizontally, vertically, or diagonally—and calls out "Bingo!"

Each number is equally likely to be called, so it seems that a winning card is just as likely to have a horizontal bingo as a vertical bingo. This would be true if there were only one player. However, the authors' mathematical analysis showed that, while horizontal and vertical bingos are equally likely on any single card, in a Bingo tournament with hundreds or thousands of cards, the first card that has a bingo is more likely to have a horizontal bingo than a vertical bingo.

This paradox is due to the fact that the 75 possible numbers are not assigned randomly to any space on the card. The first column (B) can have only the numbers 1 through 15; the second column (I) numbers 16 through 30, and so on. This structure makes it easier for players to find numbers.

B	I	N	G	O
9	29	42	60	73
6	17	41	54	75
3	21	FREE	48	62
14	19	34	55	68
11	16	31	47	65

Figure 4.2 A bingo card

If the number 14 is called, for instance, players need look at only the B column. A surprising consequence of this structure is that when there is a winning Bingo card, it is more often a horizontal win.

One way to think of this is that although horizontal and vertical bingos come up equally often, there are fewer combinations of vertical numbers that give a bingo. For example, a bingo in the B column can involve only combinations of the numbers 1 through 15, but there are far more ways to get a horizontal bingo in the first row. This means that ties are more likely with vertical bingos. In a 1000-card game, for example, horizontal and vertical bingos come up equally often, but games with horizontal winners outnumber games with vertical winners 3 to 2 because the vertical winners are more likely to occur in a cluster. When a horizontal bingo wins, there are an average of 1.24 winning cards; but when a vertical bingo wins, there are an average of 1.83 winning cards. Thus, in a given game, when "Bingo!" is called, it is more likely to be horizontal.

The mathematicians worked out the math and the coders wrote the Monte Carlo simulations. They got the same answers. The theoretical probabilities confirmed the simulations, and the simulations confirmed the theory. Good data scientists use a variety of approaches to check and recheck their results.

The Office

Some people think that computers are smarter than humans, even smarter than the developers who write the code that tell computers what to do. They think computers will soon write their own code, making human programmers obsolete. Jay's brother worked for a start-up company that made a heroic effort to build software that could go straight from requirements to solution without requiring human programmers. While visions of dollars danced in their heads, they called it an attempt to "solve the universe." It didn't.

Jay had his own experience that taught him that computer-generated code is not your friend. It also introduced him to the craziest characters and the most surreal working environment of his life—his own version of the television show *The Office*.

A large home-building company wanted a custom application that would guide homebuyers through the entire process of buying a home, from choosing custom features to the pre-approval of a loan. It was an ambitious project, but a software company persuaded them that it could be done quickly and inexpensively. Their lead programmer, George, preached of the power of computer-generated code and he was a persuasive preacher.

After a few months, it became clear that the hype was bigger than the results. Jay's company was hired to try to salvage the project and meet the now seemingly impossible deadlines.

It would be the first project for Jay's friend and coworker, a very good programmer named Bart. He was initially worried about his lack of experience, but soon realized that he was more capable than most of the original programmers. He noticed that George's coworkers would wait for George to show up before doing anything. They wouldn't code and they wouldn't answer questions. Then, when George showed up, they would do what he told them to do and refer all questions to him.

George still believed in computer-generated code, and was confident that only a few tweaks were needed to get the program working, but the days dragged on. When Bart dug deep into the code, he found bizarre bugs, such as a user's True/False choice recorded as True/True. (Inexplicably, True and False values were both considered True and a Blank value was considered False.) A human programmer would never do anything that nutty.

In addition to strange bugs in unexpected places, the code was mind-bogglingly complex, evidently intended to cover all possible scenarios. What if a customer wanted to buy a dozen houses? What if a customer wanted a thousand marble countertops in one house? A human programmer wouldn't waste time writing code to handle every preposterous possibility, but this auto-generated monster of a program did, which resulted in extremely bloated code that was difficult to debug.

Jay's team tried desperately to get the program working. Jay worked 85 hours one week, but Bart set the record, working a 98.5 hour week while, at one point, the business analyst (Jane) was shaking his chair as he worked and asking "Is it done? Is it done?" Bart never cracked. Once, during Lent when Bart abstained from sweets, someone pinned his favorite treat (a marshmallow peep) above his desk to tempt him. Bart left it there, unfazed, while he cleaned up the computer-generated code. Bart was paid an annual salary and calculated his effective hourly rate to be below that of fast-food workers. However, he took it all in stride and pushed on, repairing the code one bug at a time. He was called "the smiley guy."

Jane was the opposite. Bart was calm; Jane was frazzled. Bart abstained from cursing and use his own replacement words such as "Tokugawa!," "Toodily Do!," and "Zut Alors!" Jane freely dropped the f-bomb. It didn't take a data scientist to notice that the time when she unleashed her first f-bomb of the day was inversely related to when she had her first coffee.

Even Jay, who could usually defuse confrontations through humor, was on the receiving end of a memorable stream of obscenities when he couldn't bring himself to say that they would meet an unreasonable deadline:

Jane: Just tell me that it will be done by then.

Jay: Fine. It will be done by then.

Jane: Thank you!

(pause)

Jay: You know, just because I SAY that it will be done doesn't mean it will actually be done.

Jane: &#%!@?!!&?@#!#%&! *#%$!

Never one to give up, Jane once worked all night with George and a database guy named Erik to prepare for a demonstration of how the app would work. At around 3 a.m., George snapped. He said, "I'm sorry, I can't do it!" and left the building.

Erik worked the rest of the night with Jane on the demo. He was very capable and might have been mistaken for a machine because he took things literally. Literally. Jane once asked Erik if he could do something and he said yes. She was then surprised a week later that he hadn't even started. He told her, "You asked me if I COULD do it, not if I WOULD do it."

The morning after the all-nighter, Jane and Erik used smoke and mirrors to fake their way through the demo. Incredibly, Jane was reprimanded by her boss, not for any problems with the demo, but because she wasn't wearing the official company T-shirt. (Friday was company T-shirt day.) Jane was still wearing yesterday's clothes because she had worked on the demo all night.

Jane was well-liked within the company. When she wasn't frazzled, people teased her about a variety of things, like unknowingly adopting a strange facial tick from a coworker. And she was teased about the fact that two cat pictures on her desk looked like pictures of the same cat. She once tried to convince Jay and his wife that she had two cats by making them wait outside her apartment for proof. She ran inside, grabbed a cat, showed it to them, and yelled, "One!" Then she ran back inside, brought out another cat, and yelled, "Two!" Jay waited a few days before pointing out that she probably just showed them the same cat twice.

After George cracked, the company decided to scrap the computer-generated code entirely and start over from scratch. Jane's boss was living on antacid pills due to the stress. He knew that it would take a heroic effort to get anywhere close to meeting the deadlines, so he went on a hiring binge and the offices were overrun with programmers and consultants. Some worked out, like the Indian programmer who knew keyboard shortcuts so well that he never used a mouse and was five times faster than anyone else. Some consultants weren't so lucky:

Jane:	I'm sorry, we have to let you go.
Woman:	But you said you had wiggle-worm in the budget!
Jane:	What? Wiggle-worm?
Woman:	Yes, you told me there was wiggle-worm in the budget and that I didn't have to worry!
Jane:	Wiggle-worm?
Woman:	What?
Jane:	Did you say wiggle-worm?
Woman:	I SAID WIGGLE-ROOM!!!

A high-priced consultant named Phil was hired as the project manager. He was pretty intense and never wanted anyone to leave the office, ever. He was the kind of guy who would take a business call while his baby was being baptized. (There was a rumor that he had actually done that.)

Once, because the long hours meant that spouses rarely saw each other, a programmer's wife brought their baby triplets into the office to sit on the carpet as he worked all day Sunday. There was no air conditioning on during the weekends, so she and the babies suffered along with the software developers. At 9 p.m., she approached Phil and said, "The babies need to go home because it's bedtime." He said, "Go ahead, but your husband has to stay because he still has work to do."

The husband was a disgruntled and recalcitrant programmer who wore army boots every day. He was very smart, maybe too smart, because he truly seemed to enjoy complexity for its own sake. It seemed that this programmer had created his own guaranteed job security, because nobody else understood his code—which meant that nobody else could modify it when modifications were needed. However, the company was saddled with "technical debt" from his design decisions. On one occasion, he added entire screens that nobody asked for. On another, he proudly showed Jay that he had added a senseless button that would allow a user to increase or decrease the quantity of all home customizations simultaneously. The user could easily double the number of sinks, microwave ovens, and marble countertops—though no one would ever want to do so. His creativity and intelligence weren't enough to offset his neglect of what mattered: a robust program that met user's needs and could be modified easily. He was a programmer who worshiped computers and it turned out he didn't have job security.

Different people dealt with Phil and the pressure in different ways. One programmer, a Muay Thai kickboxer in his free time, said that, since Phil wasn't very technical, "I just blast him with BS. Blast him!" The others got a good laugh out of that one, but Phil wasn't amused when word got back to him. Another programmer, a soft-spoken guy who seemed to have everything under control, surprised everyone when they saw his bleeding knuckles. He explained "I just lost my temper and punched the wall."

Jay became good friends with a quality assurance specialist named Armando and they had their own game plan. Jay and Armando kept an eye out for Phil around lunch time and, if the coast was clear, they would literally RUN to escape the building without being seen. Armando could easily make it as a comedian with his never-ending collection of stories ("I cut my hair short so that my mom couldn't shake me by it anymore") and his infectious laughter was a lifesaver during stressful times.

The only time Jay remembers getting Phil's permission to go home was shortly after midnight on a Thursday. Jay had requested Friday off to take a programmer certification exam in the morning, so he made his carefully prepared argument: "You know, as of now, technically, it's Friday and I have Friday off. May I go home now?" Somehow, Phil was amused enough by this reasoning that he allowed Jay to leave the building without running.

With all the long hours and the missed deadlines, it sometimes seemed that the project was cursed. For example, when Jay was working on the loan pre-qualification screen, he scheduled numerous interviews with a loan officer to make sure he got the screen right, but something always seemed to come up that caused the interview to be postponed. When Jay finally got the loan officer to sit down with him in the office kitchen, they were two minutes into the interview when a power outage plunged the entire building into total darkness. They sat there for a few minutes hoping the power would come back on before finally giving up: "I guess we need to reschedule again."

The initial decision to use computer-generated code seemed like a fast, easy, and inexpensive shortcut. However, most experienced developers would have recoiled at the idea of trusting computer-generated code for such a large and important project. They know that one of the most important considerations for writing code is that it is maintainable—that programmers understand the code well enough to make the inevitable changes and bug fixes. Human programmers know that code should be written so that someone else can step in and pick up where they left off— that is, unless they want to experience firsthand a pressure-packed, over-worked, real-life sitcom like Jay did.

Computers are not going to replace developers or data scientists any-time soon.

Deep neural networks

Neural network models were first constructed in the 1940s, fell out of favor, and are again fashionable, now with more bells and whistles and the imposing name *deep neural networks* (DNNs). Most impressively, they have been used to create world-class game players: TD-Gammon was the first expert-level self-taught backgammon player and AlphaGo recently beat the world's best Go players. With these high-profile successes, it looked as if deep neural networks could be applied everywhere and would finally lead to the realization of the dream of genuine artificial intelligence.

The label *neural networks* suggests that these algorithms replicate the neural networks in human brains that connect electrically excitable cells called neurons. They don't. We have barely scratched the surface in trying to figure out how neurons receive, store, and process information, so we cannot conceivably mimic them with computers.

A neural-network algorithm is simply a statistical procedure for classify-ing inputs (such as numbers, words, pixels, or sound waves) so that these data can mapped into outputs. The process of training a neural-network model is advertised as machine learning, suggesting that neural networks function like the human mind, but neural networks estimate coefficients like other data-mining algorithms, by finding the values for which the mod-el's predictions are closest to the observed values, with no consideration of what is being modeled or whether the coefficients are sensible.

A software engineer blogged that,

No matter which part of an application of machine learning they are familiar with, whether it's computer vision or speech recognition, seasoned professionals will be able to put their experience to good use in the financial sector. At its root, deep learning has the same basics regardless of application or industry, and it should be easy for someone experienced to switch from theme to theme.

Who needs knowledge? Who needs expertise? Just hire a programmer to create a deep neural network and let it loose—it can play Go, drive cars, trade stocks, and solve the mysteries of the universe in its spare time. Computers don't need humans because they are smarter than humans.

Deep neural networks have an input layer and an output layer. In between, are "hidden layers" that process the input data by adjusting various weights in order to make the output correspond closely to what is being predicted. Data scientists describe the process using language like this:

The neural network trains itself by adjusting the weights between pairs of neurons, using a backpropagation algorithm in a combination with a stochastic gradient descent in order to minimize the loss function.

It sounds very impressive, but it is just a different kind of pattern identification. The mysterious part is not the fancy words, but that no one truly understands how the pattern recognition inside those hidden layers works. That's why they're called "hidden." They are an inscrutable black box—which is okay if you believe that computers are smarter than humans, but troubling otherwise.

Neural-network algorithms are useful for more than games—for example, for language translation and visual recognition—and they will become even more useful, but it is misleading to think that they replicate the way humans think. Neural-network algorithms do not know what they are manipulating, do not understand their results, and have no way of knowing whether the patterns they uncover are meaningful or coincidental. Nor do the programmers who write the code know exactly how they work and whether the results should be trusted. Deep neural networks are also fragile, meaning that they are sensitive to small changes and can be fooled easily.

Making sense out of words

In the 1950s, a Georgetown-IBM team demonstrated the machine translation of 60 sentences from Russian to English using a 250-word vocabulary and six grammatical rules. The lead scientist predicted that, with a larger vocabulary and more rules, translation programs would be perfected in three to five years. He had far too much faith in computers. It has now been more than 60 years and, while translation software is impressive, it is far from perfect. The hurdles are revealing.

Humans translate passages by thinking about what the author means, and then trying to convey the same meaning in another language. Translation programs do not consider content because they do not understand what words mean in any relevant sense. Computers are like New Zealander Nigel Richards, who memorized the 386,000 words in the French Scrabble Dictionary and won the French-language Scrabble World Championship twice, even though he doesn't know the meaning of the French words he spells. In the same way, translation programs identify words and phrases in an input-language sentence and rummage through a database of text translated by humans to find corresponding words and phrases in the output language, without any comprehension of what the words mean in either language.

When Gary was managing a club soccer team a few years ago, one of the parents emailed Gary that he would bring a goat to practice on Wednesday to be tested for the team. The parent did not speak English and had composed his email to Gary using a translation program that confused a goat with a child. The program was matching words in different languages without understanding the context—that a goat would not be allowed to play on a club soccer team.

Many machine translation programs, including Google Translate, now use DNNs that have improved language translation dramatically, but are still limited by the reality that, unlike human brains, computers do not truly understand words. Finding matching words in another language, and putting the matches in a grammatically correct order is *not* the same as conveying meaning. This is true not just of translation programs, but of all computer software that manipulate words. Computers find words, but do not understand ideas.

Here is a snippet from an article Gary wrote for *Fast Company* in October 2018:

Big data, it seems, knows best.... Don't buy it.... Intimidated by the algorithms, humanity could use a little pep talk.

We used Google Translate's DNN algorithm to translate Gary's passage into Spanish and back into English:

Big data, it seems, tastes better…. I did not buy it …. Intimidated by the algorithms, humanity could use a small talk.

The phrase *knows best* somehow morphed into *tastes better*. The advice *don't buy it* lost its meaning when it was turned into the statement *I did not buy it*. Finally, *pep talk* became *small talk*, two completely separate concepts. The passage's original message was completely lost because Google Translate made no attempt to understand what Gary was saying.

Here is a more puzzling example. Some researchers recently gave an AI program an essay on the 1939 American comedy *Maisie*, about a sassy showgirl who finds herself stranded in Wyoming. The word *film* was used several times in the essay and the AI program was 99.6 percent sure the essay was about a film. Then the researchers misspelled *film* one time as *flim*, and the AI program was now 99 percent certain that the essay was about a company. The researchers have no idea why this one misspelled word mattered or why the program decided this was now an essay about a company.

In July 2018, several peculiar glitches in Google Translate were discovered. When the word *dog* was typed eight times, the translation from Hawaiian to English was, "Do you want a dog to accept Jesus and be saved?" Typing *dog* 20 times and translating from Hawaiian to English yielded, "Doomsday Clock is three minutes at 12. We are experiencing characters and a dramatic developments [sic] in the world, which indicate that we are increasingly approaching the end times and Jesus' return." People with clearly too much time on their hands discovered several other doomsday messages by playing around with unusual combinations of words in several languages.

Google has no idea why this happened, because they don't know the details of how their black-box algorithm works. The best a Google spokesperson could come up with is, "This is simply a function of inputting nonsense into the system, to which nonsense is generated." The larger point is simply that Google Translate and other translation programs sometimes give bizarre results because they truly have no idea what words mean. By the time we tried translating 20 *dogs* from Hawaiian to English, Google had overridden the algorithm and all we got was 20 *dogs*. Boring! However, we tried a variation and were able to get more nonsense. Google translated *dog dog dog dog dog dog dog dog* from English to Hawaiian as *īlio'īlio 'īlio'īlio'īlio'īlio'īlio'īlio*. Then, when Google translated *īlio'īlio'īlio'īlio'īlio'īlio 'īlio'īlio* back from Hawaiian to English, we got *liver cell phone*. Go figure.

The illusion of understanding

Sometimes natural language processing (NLP) results can dazzle us. For example, following the J. R. Firth sentiment, "You shall know a word by the company it keeps," modern statistical techniques have become particularly effective at solving word analogy tasks. For example, if you trained a model on Game of Thrones, you could ask something like "Daenerys is to Drogon as Arya is to _____" and get back "Nymeria." Impressive results like that give the misleading impression that the model "understands" Game of Thrones so deeply that it can make analogies.

Being pattern-finding animals, we need to be careful not to read too much into computer output. One of Jay's projects in his NLP class at Berkeley involved analyzing a collection of 25,000 publicly available dream-journal entries. The idea was that a dream is like a "random spin through a dreamer's cognitive rolodex" and can provide insights about the dreamer.

Jay's team wrote a cutting-edge algorithm that analyzed dreams by assembling a coherent list of related words. His classmates laughed when the algorithm described the dreams from one dreamer with these words:

[belt, chicken, marketed, cheeseburger, supposedly]

It seemed a safe bet that this dreamer is concerned about weight. Eating too much chicken and too many cheeseburgers will make your belt too small. However, it turned out that the belt referred to a martial-arts belt and the dreamer was not at all obsessed with food.

It's like when you read an astrology chart and think that it matches you perfectly. Your newspaper doesn't know anything about you, but you want to believe that it does. In the same way, there is a natural tendency to believe that our computer understands us when, in fact, it understands nothing about us—or the world for that matter.

Winograd Schemas

There are many ways to show that computer programs do not understand the world because they don't know what words mean. One simple demonstration involves a test devised many years ago by Stanford computer-science professor Terry Winograd. Here is an example of these Winograd Schemas from a collection compiled by Ernest Davis, a computer-science professor at New York University:

The trophy would not fit in the brown suitcase because it was too [big/small].

Humans know that if the bracketed word is *big*, then *it* refers to the trophy, and if the bracketed word is *small*, then *it* refers to the suitcase, because, from their life experiences, humans know that it is hard to put something big into a small suitcase. Computers do not know what a *trophy* is, or a *suitcase*, or what *would not fit* means. Paraphrasing Oren Etzioni, a prominent AI researcher: how can machines take over the world when they can't even figure out what *it* refers to in a sentence?

There is a Winograd Schema Challenge with a $25,000 prize for a computer program that is 90 percent accurate. In the 2016 competition, the expected value of the score for guessing was 44 percent correct (some schemas had more than two possible answers). The highest computer score was 58 percent correct, the lowest 32 percent.

Computers do not have the wisdom humans accumulate by living life. Computers do not know the answers to simple questions like the following because they do not truly know what any of the words mean:

If I were to mix orange juice with milk, would it taste good if I added salt?
Is it safe to walk downstairs backwards if I close my eyes?

While writing this book, one of Gary's students sent him a copy of an article Gary had written that used these examples and had evidently been translated to another language and then back to English. Ironically, the garbled translation made Gary's point perfectly. For example, the first question above was translated as:

If I had been to combine orange juice with milk, would it not style excellent if I added salt?

Seeing the world through pixels

Computer programs "see" objects by analyzing pixels. For example, they are trained to associate letters of the alphabet with pixels arranged in specific ways. If a letter is written in an unusual font or is distorted, a computer program may fail where a human would not. This fragility is the basis for those little web-page access boxes, like the one shown in Figure 4.3, with weird characters called CAPTCHAs (Completely Automated Public Turing tests to tell Computers and Humans Apart) that are used to distinguish humans from computer algorithms.

Figure 4.3 Prove you are not a robot

Instead of numbers and letters, some CAPTCHAs ask the user to click on photos of cars, signs, cats and other objects which may be partly obscured or viewed from odd angles, variations that are recognized by humans but baffle computer programs.

Google provides free CAPTCHA algorithms to web developers because these algorithms give Google useful information. One of Google's projects uses optical character recognition (OCR) software to digitize a vast number of books and newspapers. When the OCR programs are puzzled by a word, it is reportedly fed into CAPTCHAs for human volunteers to decipher. Human identification of house numbers and street signs via CAPTCHAs can help Google's mapping and driverless-car projects.

The point is not that Google is using our free labor or that computers will never be able to identify objects as well as humans do. Image-recognition programs are improving all the time and will someday extremely reliable. The point is that these programs do not work like the human mind, and it is consequently misleading to call them *intelligent* in all the ways that human minds are intelligent. Even if an improved computer algorithm could reliably click on partly obscured images of street signs, it would still not know what a street sign is.

Humans see things for what they are. When we see the photo in Figure 4.4, for example, we instantly grasp its essence (the yellow color, windshield, doors, rear-view mirrors, and the *school bus* nameplate) and understand how these are related. This is obviously the front of a school bus. We know this because we have seen school buses with characteristics that are similar, even if not identical.

Computers do nothing of the sort. They are shown millions of pictures of buses, horses, stop signs, and more, and create mathematical representations of the pixels. Then, when shown a new picture, they create a mathematical representation of these pixels and look for matches in their database. The process is brittle—sometimes yielding impressive matches, other times giving hilarious mismatches.

Figure 4.4 What is this?

When we used a state-of-the-art computer program to identify the image in Figure 4.4, the DNN algorithm was 96 percent certain that the image was an airplane—perhaps because of the severely sloped windshield. It somehow overlooked the *school bus* nameplate, the multiple rear-view mirrors, and all the other evidence that a human would use to recognize this as a school bus.

Humans don't match pixels. When we see a school bus, we notice what Douglas Hofstadter calls its "skeletal essence": the shape of the bus, the color of the bus, and the placement of the windows, doors, and wheels. We don't have to look at a thousand school buses to know what they look like, nor do we need to ransack our memories trying to recall something that looks exactly like this specific object. Instead, our fantastic minds are able to grasp the essential components and understand the implications of combining these components. We are not fooled if a bus happens to be dirty, has a school name on its side, is unusually large, or is partly obscured by traffic. We still know what it is. One or two buses might be enough for us to understand the important features and, not only that, to know what buses can and cannot do. They can be driven. They can carry children.

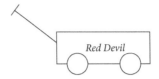

Figure 4.5 Anyone for a game of badminton?

They cannot fly. Computers do not know any of this, because they do not really know what a school bus is. They focus on the pixels and overlook the essence.

Computers can be fooled because their approach is very granular, analyzing pixels instead of concepts, and the results are sometimes preposterous. One powerful DNN program misidentified the image in Figure 4.5 as "a business." The highly touted Wolfram Image Identification Project was certain that this is a picture of a "badminton racket."

We have no idea why the Wolfram algorithm came to this conclusion and neither does anyone else. Its pixel matching may have associated the wagon handle with the handle of a badminton racket, and associated the box or the wheels with the head of a badminton racket. It certainly has no idea what a badminton racket is or what it is used for, or it would have recognized that one would not want to play badminton by swinging the object in Figure 4.5 at a shuttlecock.

Not only might a software program matching pixels in images not recognize a bus or a wagon, researchers at the Evolving Artificial Intelligence Laboratory at the University of Wyoming and Cornell University have demonstrated something even more surprising: computer algorithms may misinterpret meaningless images as real objects. For example, a powerful image-recognition program was 99 percent certain that the series of black and yellow lines in Figure 4.6 is a school bus, evidently focusing on the black and yellow pixels and completely ignoring the fact that there are no wheels, door, or windshield in the picture. Computers know nothing about buses, they just match pixel patterns.

Stop

Humans see things in context. When we are driving down a street and come to an intersection, we anticipate that there might be a stop sign. We instinctively glance where a stop sign would be placed and if we see the

Figure 4.6 An abstract school bus.

familiar red-colored, eight-sided shape with the word STOP on it, we recognize it immediately. The sign might be bent, rusty, or have a peace sign on it. We still know what it is.

Not so with image-recognition software. During its training sessions, DNN algorithms are told that the words *stop sign* go with images of many, many stop signs, so they learn to output the words *stop sign* when they input a pixel pattern that closely resembles the pixel patterns recorded during the training sessions. A self-driving car might be programmed to stop itself when it comes upon an image that it interprets as matching a pixel pattern that was labeled *stop sign* during the training sessions. However, the algorithm has no idea why it is a good idea to stop, or what might happen if it does not stop.

A human driver who sees a vandalized stop sign or a stop sign that has fallen over will stop because the human recognizes the abused sign and considers the consequences of not stopping. In contrast, because they look at individual pixels, computer programs can be led astray by trivial variations that humans know are irrelevant.

In one experiment, researchers (including a Google scientist) demonstrated that carefully selected changes in the pixels in an image of a stop sign—changes that are imperceptible to humans—caused a state-of-the-art DNN to misidentify the stop sign as a yield sign. The researchers labeled such misidentifications *adversarial attacks*, clearly recognizing the potential for mischief by troublemakers who alter stop signs in imperceptible ways to fool self-driving cars.

We did a quick-and-dirty test by putting a peace sticker on an image of a stop sign (Figure 4.7) to see what a DNN would conclude. The DNN misidentified the image as a first aid kit—perhaps because of the abundance of red and white colors.

This particular misidentification problem is well known and can have serious consequences if self-driving cars don't know when to stop and when to go. A workaround is for self-driving cars to use GPS systems and digital maps created by humans to identify permanently fixed objects, including intersections that have stop signs.

Good data scientists are looking for ways to thwart malevolent people who might launch adversarial attacks on all sorts of sights and sounds used by AI algorithms. In January 2018 the organizers of a machine-learning conference to be held in April announced that they had accepted 11 papers proposing ways to thwart adversarial attacks. Three days later, an MIT graduate student, a Berkeley graduate student, and a Berkeley professor reported that they had found ways to work around 7 of these defense systems. At the April conference, they won a best paper award. There is clearly an AI arms race.

Figure 4.7 Stop or you may need first aid

Computers are useful, not omniscient

Computers are hardly useless. To the contrary, computers are extremely useful for the many specific tasks that they have been designed to do, and they will get even better, much better. However, we should not assume that computers are smarter than us just because they can tell us the first 2000 digits of *pi* or show us a street map of every city in the world.

One of the paradoxical things about computers is that they can excel at things that humans consider difficult (like calculating square roots) while failing at things that humans consider easy (like recognizing stop signs). They do not understand the world the way humans do. They have neither common sense nor wisdom. They are our tools, not our masters.

Good data scientists know that data analysis still requires expert knowledge.

The Fourth Pitfall of Data Science is:

Worshiping Computers

Torturing Data

> If you torture data long enough, it will confess.
> —Ronald Coase

It has been estimated that 40 percent of all adults in the United States are on diets, though they drift in and out, averaging four-to-five diet attempts a year, while they spend $40 billion annually on books, videos, surgeries, and gimmicks, looking for ways to lose weight easily.

The truth that people don't want to hear is that the most reliable way to lose weight is to exercise more and eat less. Period.

Exercise regularly, that's no fun. Eat less, that's definitely no fun. Enter the hucksters who market weight-loss scams, including creams that melt away fat, sunglasses that make food look yucky, and magnetic earrings that stimulate acupressure points.

One weight-loss guru, Brian Wansink, had some very impressive academic credentials. He was a Professor of Marketing at Cornell and the Director of the Cornell Food and Brand Lab. He authored (or coauthored) more than 200 peer-reviewed papers and wrote two popular books, *Mindless Eating* and *Slim by Design*, which have been translated into more than 25 languages. His work was reported in the *Oprah* magazine, *Time* magazine, *USA Today*, *New York Times*, and on the *Today* show.

In one of his most famous studies, 54 volunteers were served tomato soup. Half ate from normal bowls and half ate from "bottomless bowls" which had hidden tubes that imperceptibly refilled the bowls. Those who ate from the self-refilling bowls ate, on average, 73 percent more soup,

The 9 Pitfalls of Data Science. Gary Smith and Jay Cordes. Oxford University Press (2019). © Gary Smith and Jay Cordes 2019. DOI: 10.1093/oso/9780198844396.001.0001

though they did not report feeling any fuller than the people who ate from normal bowls. Wansink and his coauthors concluded that, "the amount of the soup remaining in the bowl provides a visual cue that indicates whether he or she should continue eating or should stop." Eating is evidently not about filling a stomach, but about emptying a bowl.

Wansink was given an Ig Nobel award in 2007 for his bottomless-bowl study. Another Ig Nobel winner that year was for a study reporting that rats cannot always tell the difference between someone speaking Japanese backwards or Dutch backwards.

But many people took the bottomless-bowl study seriously, including Wansink. In other studies, he reported that people eat more when they use bigger plates, that people at Super Bowl parties eat more when the snacks are put in larger bowls, and that people eat more popcorn, no matter whether it is fresh or stale, if they are given bigger cartons of it. He was an advocate for 100-calorie snack packs, reasoning that eating is not about filling a stomach, but emptying a snack pack. These strategies all had the allure of being painless ways to lose weight.

Wansink reported many other simple tricks. People who leave cereal boxes in plain view weigh more, so hide the cereal. Kids whose parents tell them to clean their plates ask for more food, so stop nagging. Kids who order their school lunches ahead of time are less tempted to buy junk food, so order ahead. People who watch TV during dinner eat more, as do people who watch "distracting" shows, so turn off the TV or watch something that you don't want to watch.

So many easy ways to lose weight! No wonder his studies were so widely reported.

In 2016 the trouble started. In a blog post titled, "The Grad Student Who Never Said No," Wansink wrote that,

With field studies, hypotheses usually don't "come out" on the first data run. But instead of dropping the study, a person contributes more to science by figuring out when the hypo[thesis] worked and when it didn't. This is Plan B. Perhaps your hypo[thesis] worked during lunches but not dinners, or with small groups but not large groups.

This sounds an awful lot like data mining.

Wansink gave an example of a PhD student from a Turkish university who came to work in Wansink's lab in 2013 and was given data that had been collected at an all-you-can-eat Italian buffet:

When she arrived, I gave her a data set of a self-funded, failed study which had null results . . .. I said, "This cost us a lot of time and our own money to collect. There's got to be something here we can salvage because it's a cool (rich & unique) data set." I had three ideas for potential Plan B, C, & D directions (since Plan A had failed) . . .

Every day she came back with puzzling new results, and every day we would scratch our heads, ask "Why," and come up with another way to reanalyze the data with yet another set of plausible hypotheses. Eventually we started discovering solutions.

By never saying no, the student got four papers (now known as the "pizza papers") published with Wansink as a coauthor. The most famous one reported that men eat 93 percent more pizza when they eat with women.

One researcher commented on the blog: "Brian - Is this a tongue-in-cheek satire of the academic process or are you serious?" Others took a closer look at the pizza papers and then started scrutinizing other papers. They found, in addition to enthusiastic data mining, several data inconsistencies (like numbers that didn't add up) and inappropriate statistical tests. Stunned by the damaging publicity, a faculty committee at Cornell University began investigating.

Email correspondence with the Turkish graduate student surfaced in which Wansink advised her to separate the diners into "males, females, lunch goers, dinner goers, people sitting alone, people eating with groups of 2, people eating in groups of 2+, people who order alcohol, people who order soft drinks, people who sit close to buffet, people who sit far away, and so on." Then she could look at different ways in which these subgroups might differ: "# pieces of pizza, # trips, fill level of plate, did they get dessert, did they order a drink, and so on." He concluded that she should, "Work hard, squeeze some blood out of this rock." She responded, "I will try to dig out the data in the way you described."

A Cornell student who had worked as an intern in Wansink's lab said, "I remember him saying it so clearly: 'Just keep messing with the data until you find something.'" She was so uncomfortable with this directive that she left the lab before her internship ended.

Brian Nosek, a crusader against p-hacking, called Wansink's research a "caricature" of p-hacking. Soon, critics had created a "Wansink Dossier," listing errors, inconsistencies, and dubious practices in his research. After a year studying Wansink's research, Nick Brown, a graduate student at the University of Groningen in the Netherlands, concluded that, "They're doing the p-hacking and they're getting other stuff wrong, badly wrong. The level of incompetence that would be required for that is just staggering."

In September 2018, the Cornell faculty committee investigating Wansink concluded that he had "committed academic misconduct in his research." Wansink resigned, effective the following June, and "has been removed from all teaching and research. Instead, he will be obligated to spend his time cooperating with the university in its ongoing review of his prior research."

More than a dozen of Wansink's published research papers have been retracted, including the papers about clean plates, pre-ordered lunches, Super Bowl snack bowls, distracting TV, and one of the pizza papers. Good data scientists avoid embarrassing retractions by doing it right the first time: "Check yourself before you wreck yourself."

Fishing expeditions

Researchers seeking fame and funding may be tempted to do what Wansink did—torture the data to find novel, provocative results that will be picked up by the popular media. Tweets can be used to predict stock prices. Hurricanes with female names are deadlier. People with positive initials live longer than do people with negative initials. Asian Americans are prone to heart attacks on the fourth day of the month. Dead fish display brain activity when shown photographs.

Provocative findings are provocative because they are novel and unexpected, and they are often novel and unexpected because they are simply not true. So, research that gets reported in the popular media is often wrong—which either fools people or undermines the credibility of scientific research.

Several years before the Wansink scandal, Daryl Bem, a prominent social psychologist, offered researchers this advice:

Examine [the data] from every angle. Analyze the sexes separately. Make up new composite indexes. If a datum suggests a new hypothesis, try to find further evidence for it elsewhere in the data. If you see dim traces of interesting patterns, try to reorganize the data to bring them into bolder relief. If there are participants you don't like, or trials, observers, or interviewers who gave you anomalous results, place them aside temporarily and see if any coherent patterns emerge. Go on a fishing expedition for something—anything—interesting.

Using the fishing-expedition approach, Bem was able to discover evidence for some truly incredible claims, such as "retroactive recall"—people are more likely to remember words during a recall test if they study the words *after* they take the test. For example, if they manipulate the word *tree* after taking a recall test, this increased the probability that they remembered the word *tree* during the recall test.

As might be expected, this was complete nonsense, and other researchers could not replicate Bem's fishing expedition. In the famous words of Ronald Coase, "If you torture data long enough, it will confess."

Good data scientists do not seek false confessions.

Flipping ten heads in a row

Derren Brown is a mentalist who says that he can flip ten heads in a row with a fair coin. This is an astonishing claim since there is only a 1 in 1024 chance of flipping ten heads in a row by luck alone. Brown backed up his claim with a video filmed from two angles. There were no cuts in the video, it wasn't a trick coin, and there were no magnets or other trickery involved. In a later video, he gave away his secret: he had simply filmed himself flipping coins for nine hours until he got ten heads in a row. The video seemed magical, but it was a tedious trick.

The posted video is just a joke, but Brown's coin-flip prank is a clear example of the survivorship bias that occurs when our perception of data is distorted by what Nassim Taleb called the "silent evidence" of failures. If we don't know about the failures, how can we evaluate the successes?

P-hacking

Statistical significance is an odd religion that researchers worship almost blindly. In Chapter 3, we discussed a study of whether regular doses of aspirin reduced the risk of a heart attack. The subjects were divided randomly and either took an aspirin tablet every other day or took a placebo. When the data were collected, the statisticians moved in. The statistical issue is the probability that, by chance alone, the difference between the two groups would be as large as the difference that was actually observed. Most researchers consider a probability (p-value) less than 0.05 to be "statistically significant." In other words, patterns in the data are considered statistically persuasive if they have less than a 1-in-20 chance of occurring by luck alone.

The results from the aspirin study were highly statistically significant. The probability that the observed differences would be so large by luck alone was 0.0067 for fatal heart attacks and 0.0000105 for nonfatal heart attacks.

Most experiments are not so successful and do not yield statistically significant results. Nonetheless, a study of psychology journals found that 97 percent of all published test results were statistically significant. Surely, 97 percent of all the tests that were conducted did not yield statistically significant results, but editors generally believe that tests are not worth reporting unless the results are statistically significant. This bias means that published studies are disturbingly similar to the video of the mentalist flipping heads—we see the successes, but we do not see the silent evidence of failures.

P-hacking refers to the practice of testing many theories, but only reporting the results with low p-values. Even if every theory tested by a hapless researcher was worthless, we can expect one out of twenty to be statistically significant, and potentially publishable. Now imagine that hundreds of researchers test thousands of worthless theories, write up the statistically significant results, and discard the rest. The problem for society is that we only see the tip of this statistical iceberg. If we knew that behind the reported tests were thousands of unreported tests and remember that, on average, one out of every twenty tests of worthless theories will be statistically significant, we would surely view what does get published with considerable, well-deserved skepticism.

Pharmaceutical companies, for example, test thousands of experimental drugs and, even with well-designed, unbiased studies, we can expect hundreds of worthless drugs to show statistically significant benefits ("false positives") which can generate immense profits. Drug companies have a powerful incentive to p-hack, and there is little incentive for them to retest an approved treatment to see whether the initial results were just a fluke.

If the video of Derren Brown flipping coins were presented as evidence of his supernatural control of coins, he could claim that the results were statistically significant with a p-value of 0.001. However, since he tried for many hours and hid the video record of his failures, he was guilty of p-hacking. The true p-value is 1, because he flipped coins until he got the results he wanted.

If there are multiple tests and selective reporting, p-values are not only meaningless, but misleading. And yet that is exactly what data scientists do when they use a computer algorithm to discover statistically significant patterns in the data and pay no attention to all the silent failures.

P-hacking is particularly insidious because it cannot be detected easily. The evidence has been destroyed and the only indication that a sin was committed is found afterward, if someone else tries to replicate the results and fails. Unfortunately, most researchers do not spend their time trying to replicate other people's studies because reputations are built by discovering new results, not by confirming old ones. P-hacking is the perfect crime.

Bem's advice to researchers covers most of the ways to p-hack: look at many variables, look at different measures or indexes, look at different possible relationships, separate the data into subgroups, discard some data—go on a fishing expedition. For serious data scientists, reporting the results of a fishing expedition is a sin, not a recommendation.

The common thread through these nefarious methods is that we are misled by the data we *don't* see. If we watch a video of someone making a circus basketball shot where the ball bounces off a wall, a car, and a roof before going into the basket, we know that hours of video footage didn't make the cut, because we know how hard it is to make circus shots. However, when we see a published study with a semi-plausible result— like eating less off smaller plates—we might not suspect a trash can full of unreported failures. The only way to know for sure is to redo the study. When researchers try to replicate published studies, they often fail. These failures are so common that this pattern is called the *replication crisis*.

Tweets and heart attacks

A 2015 study examined a random sample of 826 million Twitter tweets collected between June 2009 and March 2010. Using locations given in some of the user profiles, the researchers were able to link 148 million tweets to 1347 counties across the United States. After analyzing these data, they concluded that counties in which tweeters frequently used "negative language" tended to have higher mortality rates from athero-sclerotic heart disease (AHD) than did counties in which tweeters were more inclined to use "positive" words.

This study suggests that Twitter users who are upset, depressed, or angry are at risk for fatal heart attacks. As appealing as that conclusion may be, Nick Brown and James Coyne pointed out several problems with this study. For one, this analysis does not tell us whether an *individual* who is prone to negative tweets is more likely to suffer a fatal heart attack. Instead, it looks at countywide data on tweets and heart attacks.

This focus on counties rather than individuals might be justified if people are homogeneous within counties, and different across counties. Perhaps some counties are hotbeds of negativity, and this toxic environment is deadly. However, counties are not homogeneous. Just look at the counties that contain large cities like Los Angeles, New York, Chicago, and Houston.

In addition, the authors tortured the tweets. They initially found that positive words were *positively* correlated with heart attacks, the opposite of what they expected, so they removed *love* from their list of positive words in order to get the negative correlation they wanted! Surely, *love* is near the top of any list of positive words, and it is wrong to pretend otherwise.

As a robustness check, Brown and Coyne looked at how well the authors' positive and negative words predicted suicides. If negative words are so upsetting that they trigger heart attacks, shouldn't they also trigger thoughts of suicide? They checked and found that counties with a plethora of negative words had fewer suicides! Do crude outbursts trigger heart attacks, but protect people from self-harm? Maybe we should remove some more words and flip the results? Or maybe this is nonsense.

Nutritional studies

Are eggs healthy? How about coffee? Does wine cure cancer or does it cause cancer? The seemingly constant reversals in nutritional advice is maddening. Sometimes it seems like science is broken. One explanation is that scientists cannot always do randomized controlled trials (RCTs), like the aspirin/heart-attack study, where researchers compared randomly selected treatment and control groups.

Randomized controlled trials are usually impractical for assessing dietary advice. Imagine the difficulty in randomly assigning people to either drink or not drink coffee and then ensuring that they stick to their assigned regimen for decades while their health is monitored. Unbelievably, such a study was once done, but only because a Swedish king ordered that it be done.

In the 1700s King Gustav III wanted to prove that coffee was a slow-acting poison, so he offered a deal to identical male twins who were waiting to be beheaded for committing a murder. Their sentences would be changed to life imprisonment if one of them drank three pots of coffee every day for the rest of his life, while his twin drank three pots of tea.

Gustav thought that he had condemned the coffee-drinking twin to death, though more slowly than with a beheading, and that his coffee-induced death that would prove that coffee was poison.

Both twins outlived Gustav (who was assassinated), and the coffee-drinking twin outlived the tea drinker. The choice of identical male twins was a clever way of avoiding the confounding effects of gender, age, and genes. However, we now know that nothing statistically persuasive can come from a sample with two subjects.

These days, researchers are not allowed to sentence people to a life of coffee drinking or coffee abstinence, so they rely instead on observational data. They identify people who drink coffee and people who do not, and then compare the health of the two groups. They try to control for possible confounding factors, but may overlook some important ones. For a while, the evidence seemed to be that coffee causes a variety of medical problems. However, most of these studies were flawed and it is now believed that coffee increases life expectancy.

Why do researchers keep changing their minds? Coffee was bad, now coffee is good. Chocolate was bad, now chocolate is good. Wine was bad, now wine is good. Sunshine was bad, now sunshine is good. Vitamin D supplements are good; now they are worthless. The recommendations keep changing because confounding factors were ignored or because the data were ransacked to find something publishable. Data scientists should not data mine or torture data, and they should be wary of observational data.

A new study shows . . .

A legendary hedge-fund trader named John Arnold retired as a billionaire at the age of 38 and shifted his focus to philanthropy. In 2008, he and his wife started the Laura and John Arnold Foundation to tackle, among other things, the replication crisis. John Arnold had become skeptical of practically every new scientific finding and once tweeted that he considered "a new study shows . . .." " as "the four most dangerous words." The Arnolds decided to identify and support reformers, including Brian Nosek, who we encountered earlier in this chapter, and John Ioannidis, who we will encounter in the next chapter.

Nosek had started a Reproducibility Project which would attempt to replicate 50 psychology studies to see how many could be verified. With the Arnold Foundation's backing, Nosek and Jeffrey Spies founded the

Center for Open Science, with a mission to "increase the openness, integrity, and reproducibility of scientific research." With the additional funding, the center tried to replicate 100 studies. Only 36 percent could be verified. Even among the few that could be replicated, the effects were generally smaller in the follow-up study than in the original reported results.

Some of these disappointing results can be attributed to regression to the mean (which we will discuss in Chapter 8), but the bigger problem is p-hacking. A 2018 survey of 390 professional statisticians who did statistical consulting for medical researchers found that more than half had been asked to do things they considered severe violations of good statistical practice, including conducting multiple tests after examining the data and misrepresenting after-the-fact tests as theories that had been conceived before looking at the data.

Even if researchers don't p-hack individually, there is collective p-hacking. Suppose 100 researchers test 100 worthless theories. We expect five, by luck alone, to have statistically significant results that might be accepted by journals. The other 95 tests disappear without a trace—the same as if one researcher tested 100 worthless theories and reported only the five best results. This is called the *publication effect*, in that statistically significant results find their way into journals and books, while insignificant results are not reported. It is also called the *file drawer effect*. The credibility of published studies are undermined by the studies that failed and were filed and forgotten.

Nature, one of the very best scientific journals, surveyed 1500 scientists about the replication crisis. More than 70 percent reported that they had tried and failed to reproduce another scientist's experiment, and more than half had tried and failed to reproduce some of their own studies!

Nosek worked with the Association for Psychological Science to introduce three "Open Practice Badges" (Open Data Badge, Open Materials Badge, and Preregistered Badge) for papers accepted by participating journals. Authors who meet the requirements for transparency have the badges displayed at the beginning of their articles. For example, the Preregistered Badge requires the following practices:

- The author prepares a "registered" plan for analyzing the data before examining the data.
- The registered plan is filed online with open access.
- Any deviations from the plan are reported.
- All analyses described in the registered plan are reported in the article.

The badges are a creative way of providing incentives for psychology researchers to do studies that yield worthwhile results.

The idea of pre-registering research plans is becoming more widespread. One study of heart-disease research found that, before a registry was mandated, half of the published studies had statistically significant results; afterward, only 8 percent did—barely more than the 5 percent we expect by luck alone.

All data scientists should try to follow these guidelines, even if they are not looking to publish journal articles. Otherwise, they will have their own replication crisis when their conclusions and recommendations flop, and they are exposed as data clowns.

Robo-Tester

BuyNow, a company that managed over a million Internet domains, hired WhatWorks to determine an optimal page layout. WhatWorks ran a scientifically valid experiment using a random-event generator to send customers to pages with different layouts, in order to determine which layout produced the most revenue per visitor.

WhatWorks ran its tests and gave BuyNow a report on the optimal design. BuyNow liked the report, but felt, not unreasonably, that different people visit different domains and might well have different tastes. Instead of one layout for all domains, it would be better to have custom layouts for each domain.

In theory, that made sense, but it wasn't practical for humans to run a million experiments. However, BuyNow knew that WhatWorks had developed Robo-Tester, an automated AI system for testing designs. One of its popular features was that as it accumulated evidence that one design was better than others, it automatically sent more users to that design, in order to increase revenue while it continued gathering data.

BuyNow loved the idea of Robo-Tester, but WhatWorks' data scientists were not infatuated:

WhatWorks: The domains don't have enough traffic to draw meaningful conclusions.
BuyNow: But Robo-Tester only chooses layouts when there is statistical significance.
WhatWorks: Statistical significance doesn't always mean it's significant.
(stunned silence)

WhatWorks:	What I mean is, with a million tests you'll get statistical significance all over the place due to chance. It won't mean that you found the best layouts.
BuyNow:	Yeah, you always say that.
WhatWorks:	And I'm always right!

WhatWorks' data scientists had created Robo-Tester and they recognized its limitations. One problem was that some domains had so little web traffic that it could take months, or even years, to reach a decision; meanwhile, a substantial amount of web traffic would be sent to less profitable designs during this misguided experiment.

Even if some sites might do better with a custom layout, Robo-Tester is essentially an automated p-hacker. By chance alone, thousands of ineffective layouts would be chosen and, in every case, revenue would be lost until Robo-Tester accumulated enough data to correct its mistake. Robo-Tester's replication crisis would be an expensive lesson in the dangers of p-hacking.

In addition, Robo-Tester was never 100 percent certain that it had found the best design. Even if there were an optimal design and Robo-Tester happened to find it, it would still send visitors to other layouts to gather more data. Since these domains typically had very little traffic, the results would be wildly volatile, and Robo-Tester would keep giving second and third chances to inferior layouts. The cumulative cost of the unending experimentation would almost certainly outweigh any potential benefits.

To settle the debate, an experiment was run that pitted Robo-Tester head-to-head against the single layout initially recommended by WhatWorks. The net result was that Robo-Tester reduced BuyNow's revenue by about 1 percent.

The WhatWorks data scientists were right, but the idea that automating experiments is the easy road to riches never went away.

Aspartame doesn't cause cancer

Experts are unanimous: aspartame is fine to use as a substitute for sugar. According to the U.S. Food and Drug Administration (FDA):

Considering results from the large number of studies on aspartame's safety, including five previously conducted negative chronic carcinogenicity studies, a recently reported large epidemiology study with negative associations between the use of aspartame and the occurrence of tumors, and negative findings from a series of three transgenic mouse assays, FDA finds no reason to alter its previous conclusion that aspartame is safe as a general purpose sweetener in food.

Similarly, the European Food Safety Authority reported:

On the basis of all the evidence currently available . . . there is no indication of any genotoxic or carcinogenic potential of aspartame and that there is no reason to revise the previously established ADI for aspartame of 40 mg/kg [of body weight].

In the largest study to date, the National Cancer Institute looked at over 500,000 adults and compared the ones who drank aspartame-sweetened beverages to those who didn't and found no increase in lymphomas, leukemias, or brain tumors.

So why are we still talking about this? Well, possibly because of fears inflamed by hoax emails, aspartame keeps getting tested. According to FDA officials, aspartame is "one of the most thoroughly tested and studied food additives the agency has ever approved." Over-testing is essentially p-hacking; the more studies that are done, the more likely we are to find a meaningless anomaly. Sure enough, 15 years after its approval by the FDA, an article was published that warned of a possible link between brain tumors and aspartame.

Even if research is done rigorously, it is not surprising that an occasional study will be an outlier. Just as hundreds of studies of a worthless treatment might occasionally find statistical significance, so hundreds of studies of a harmless additive might occasionally find statistical significance—collective p-hacking.

We have other reasons to be skeptical of this outlier study, most convincingly articulated in an editorial by molecular epidemiology expert Julie A. Ross. The worrisome study simply showed a correlation between an increasing number of brain tumors in the overall population and increasing use of aspartame, and offered no evidence that they were related. The study did not show that people who consumed aspartame were more likely to have brain tumors.

In the same way, the Pizza Principle says that since the 1960s, the cost of a New York City subway ride has been roughly equal to the cost of a slice of pizza. How are subway prices related to pizza prices? They aren't, but they have both increased over time.

ESP-hacking

The term *extrasensory perception* (ESP) is used to describe the purported ability to receive information without using the five known physical senses: sight, hearing, taste, smell, and touch. It is sometimes called a *sixth sense*. Extrasensory perception includes telepathy (reading another person's mind) and clairvoyance (identifying an unseen object).

J. B. Rhine is the most famous ESP researcher in history. Beginning in 1927, Rhine conducted decades of research on extrasensory perception in a parapsychology laboratory at Duke University. His most famous experiments involved a Zener pack of twenty-five cards—five cards of five designs: circle, cross, wavy lines, square, and star.

In a typical experiment, a *sender* turns the twenty-five cards over, one by one, stares at the card for a few seconds, and records the symbol, while the *receiver* writes down a guess. Rhine's 1934 book, *Extra-Sensory Perception*, described several successful experiments, and received a glowing review from the *New York Times* science editor:

[T]here can be no doubt of the value of his work . . .. Dr. Rhine has made a name for himself because of his originality, his scrupulous objectivity, his strict adherence to the scientific method.

Rhine became famous and ESP was thought to be real. People bought Zener cards and did ESP tests at home. Those who succeeded sent letters to Rhine reporting their successes. Those who failed kept their failures to themselves. It turns out that Rhine was doing the same thing—writing about his successful experiments while literally putting the failures in file cabinets.

Rhine's ESP experiments are a classic example of p-hacking. Even if there is no such thing as ESP, we expect one out of every twenty people tested—either in Rhine's lab or at home—to make enough correct guesses to be considered statistically significant evidence. With thousands of tests, some incredible results are inevitable—and these remarkable results are the ones that get reported. This selective reporting is an example of the publication effect, or file drawer effect, in that statistically significant results are publicized, while insignificant results are filed away and forgotten.

This selective reporting is the first Texas Sharpshooter Fallacy. Rhine also committed the second Texas Sharpshooter Fallacy by discovering patterns after the fact and making up explanations. If a receiver's guesses matched previous cards, this was "backward displacement;" if they matched future cards, this was "forward displacement." If there were fewer matches than expected, this was evidence of "negative ESP."

Rhine's experiments were not persuasive evidence of ESP but, instead, compelling evidence that he fell into the data-torturing pitfall.

The One Million Dollar Paranormal Challenge

The One Million Dollar Paranormal Challenge, which ran for 50 years, was a million-dollar bet that no one possesses paranormal or supernatural abilities that have no known scientific explanation. It was a wager that those who claim such abilities are charlatans—often magicians—who prey on the gullibility of those who don't know how the tricks are done.

The James Randi Educational Foundation put up $1 million dollars (the "Randi Prize") and required claims to be scientifically tested, with objective success criteria. Participants agreed ahead of time to the conditions of the experiment and then gave it their best shot.

One tried to identify playing cards in a sealed envelope, one attempted to distinguish a performance-enhancing bracelet from a decoy, one endeavored to identify objects remotely, and one tried to send electricity through his hand that could be felt by another person. They all failed.

They all made the right decision to try for the prize, even if they knew they were frauds, because they might have gotten lucky. Suppose you claim to be able to identify a playing card in a sealed envelope while wearing a blindfold. Just guess a random card and hope for the best! Even if you have to do it twice in a row to show true psychic superpowers, you have a 1/2704 chance of winning a million dollars.

You might even consider coordinating with hundreds of fake-psychic friends to p-hack the Randi Prize. Randi is very clever and probably would see through the ploy, but you never know.

A good friend of Jay's, a math professor named Don, told Jay that he was once playing a card game with a friend and noticed that they had settled into a predictable pattern in the shuffling and dealing of cards. Don wondered if his inattentive friend would notice if he peeked at the top card on the deck. Don blatantly picked up the top card, looked at it, and returned it to the deck while his friend was focused on his own cards. Don waited a few seconds and then pointed to the top card and said "8 of clubs!" Sure enough, it was. Don's friend stared at him in disbelief, trying to understand what had just happened.

Don just silently looked back at him with raised eyebrows. The game continued and, a little later, Don did it again. Not in a sneaky way, he just peeked at the top card, as if this were a normal part of the game. Then, he pointed at the card and proclaimed "9 of diamonds!" before it was flipped over and revealed. This time, the friend stopped and looked at Don for a very long time while Don did all he could to keep from laughing.

Don decided it might be more impressive if he guessed one wrong so that his friend would think that he was just blindly guessing. So, the next time, he didn't peek, pointed at the top card, and said, "king of hearts!" His friend flipped over the card and it was the king of hearts. Don finally lost it and rolled on the floor in laughter. He laughed again later when Jay pointed out that his friend was probably watching him like a hawk before his last guess.

Given enough tries, anything that can happen eventually will happen, even without supernatural abilities. Why not try to win a million dollars or at least get a good laugh out of trying?

Data abuse

When it comes to data science, however, there's nothing funny about torturing data. If you p-hack, you won't get a million dollar prize. When people realize that you are a data clown, they might suggest that you get a job in a circus.

The Fifth Pitfall of Data Science is:

Torturing Data

Fooling Yourself

A lot of evidence is very tenuous. We need to train people to understand what the limitations are, what the caveats are, how much they
can trust or distrust what they read or what they see, and what they
are being called to do. Then make them ask for better evidence.
—John Ioannidis

Dr. John Ioannidis (yo-a-NEE-dees) was walking in Palo Alto, California,
one day in the fall of 2014 when he noticed an unusual company name on
a building: Theranos. He didn't know what it meant, but he was intrigued
by the similarity to the Greek word thanatos, which means death. He
hoped that was not what they were selling! Later, he Googled the company and found out that its name was created by combining "therapy" and
"diagnosis," and that they were a highly hyped biotech start-up seeking to
"disrupt the medical diagnostic test market." Theranos claimed that it had
developed a blood-testing device named Edison that could run hundreds
of tests quickly and cheaply using a single drop of blood. People would
soon be able to have their blood tested in supermarkets or pharmacies
while they shop.

Ioannidis, the quintessential scientist, was intrigued and dug deeper,
but he couldn't find anything about how the technology worked because
Theranos's scientists didn't communicate with other scientists or publish
research documenting their product's effectiveness. Ioannidis also felt
uneasy about the idea of doing hundreds of blood tests. That's p-hacking,
because some tests, by luck alone, are bound to produce anomalous results
even for perfectly healthy people. Millions of people doing hundreds of

The 9 Pitfalls of Data Science. Gary Smith and Jay Cordes. Oxford University Press (2019).
© Gary Smith and Jay Cordes 2019. DOI: 10.1093/oso/9780198844396.001.0001

tests will surely create lots of false alarms, unnecessary worry, expensive doctor visits, and an occasional tragedy.

Ioannidis told us that, "They wanted to do something we know is a bad idea; testing multiple times will find something abnormal. People with incidentalomas can die." (Yes, that's a real medical term; it means "an incidentally discovered mass or lesion.") Ioannidis recounted the story of a 50-year-old woman who had a CT scan that showed a suspicious mass in her pancreas. Worried that it might be malignant, it was surgically removed. It turned out that the lump was not malignant, but the patient got an infection from the surgery and died. False diagnoses are not always harmless.

On the Stanford professor profile page, Ioannidis is extremely modest: "I love to be constantly reminded that I know next to nothing," which is pretty close to the Socratic Paradox, "I know that I know nothing." His accomplishments and awards are stunning: National Award of the Greek Mathematical Society, top rank of his medical-school class, honorary doctorates, and, perhaps most importantly, co-Director of the Meta-Research Innovation Center at Stanford, which is devoted to improving the quality of scientific research in biomedicine and other fields. He has been called "the godfather to the science reform crowd" and, according to the *Atlantic*, he "may be one of the most influential scientists alive."

Theranos probably now wishes they had chosen a company name that Ioannidis would not have noticed. Within a few weeks he wrote a Viewpoint which he submitted for review to the *Journal of the American Medical Association*. His paper was published in February 2015 with the title "Stealth Research—Is Biomedical Innovation Happening Outside the Peer-Reviewed Literature?" Ioannidis criticized Theranos for its stealth research—making claims without any peer review by outside scientists. Ioannidis asked, "How can the validity of the claims be assessed if the evidence is not within reach of other scientists to evaluate and scrutinize?" Soon, others were asking the same question. Where's the proof?

While the Theranos CEO was being featured on magazine covers and the media was lavishing praise on the billion-dollar company, Ioannidis had planted the seed of skepticism: "We have been misled many times about innovations in medicine before."

A lawyer from Theranos contacted Ioannidis in an effort to persuade him to pull back. They suggested that he coauthor an article with the company CEO supporting her view that an FDA approval would be the best possible evidence that their technology worked. He didn't bite. Although he is soft spoken and friendly, Ioannidis has strong beliefs backed by a lifetime of research and this wasn't the first time he'd met resistance.

Later that year, in October 2015, the *Wall Street Journal* published an investigative article on Theranos which suggested a violation of federal rules for laboratories and a manipulation of the proficiency-testing process. Perhaps more damning, they found large differences between the results of a Theranos blood test and hospital blood tests for an Arizona woman. The *Journal* became the newest target for Theranos's lawyers, but now there were too many critics to silence.

By the end of the year, the federal Centers for Medicare & Medicaid Services (CMS) stated that testing at a Theranos lab posed "immediate jeopardy to patient health and safety." The house of cards fell quickly. Theranos, once valued at $9 billion, collapsed. Their CEO, once the youngest female self-made billionaire on the planet, lost it all and faced criminal fraud charges.

The takeaway most people get from the epic failure of Theranos is that we should be more vigilant about corporate fraud. Ioannidis has a different take: "Fraud and sociopaths are the exceptions. It's people who are not well trained, using harmful research practices that we need to worry about."

The best scientists have mastered the scientific method and are responsible for the many great success stories in medical science: the HIV triple drug combination that has transformed HIV from a death sentence to a chronic condition, the benefit of statins, the effectiveness of antibiotics, and the treatment of diabetes with insulin. Many deadly diseases now have effective treatments. There are also many successes in identifying the causes of diseases: asbestos can lead to mesothelioma, benzene can cause cancer, and, of course, the association between smoking and cancer. Ioannidis praises the methodology followed in air pollution studies and holds them up as a template for others who try to answer difficult, complex questions that are not as obvious as smoking causing lung cancer:

The Six Cities and American Cancer Society studies are exemplary large-scale investigations, with careful application of methods, detailed scrutiny of measurements, replication of findings, and, importantly, detailed re-analysis of results and assessment of their robustness by entirely independent investigators.... It would be wonderful, if in the future the same rigorous re-analysis and replication standards could become the standard for all important areas of research that can inform policy.

Ioannidis is cautiously hopeful: "I think that several institutions are slowly recognizing the need to shift back to putting more emphasis on methods and how to make a scientist better equipped in study design, understanding biases, in realizing the machinery of research rather than the technical machinery."

Ioannidis has focused his career on "meta-research," or research on research. Combing through published papers, trying to replicate them, and identifying the biases in order to help him and his colleagues understand and help mitigate the replication crisis. He described the revolution he helped lead: "now everybody says we need replication; we need reproducibility. Otherwise our field is built on thin air."

When asked to identify a few rules that he'd recommend for the next generation of scientists, a few came to mind immediately:

> Think ahead of time. Don't just jump into an idea; anticipate the disasters.
> Don't fool yourself. Be skeptical of findings that agree with your expectations. If they look too good to be true, they probably are.
> Do experiments. Randomize whenever possible. Observational studies are tricky and having lots of data doesn't ensure that they're valid.

When Jay mentioned that one of our data science pitfalls is fooling yourself, Ioannidis caught the nod to physicist Richard Feynman's timeless advice: "The first principal is that you must not fool yourself—and you are the easiest person to fool."

Ioannidis has compiled a list of real-world "proven" treatments that turned out to be ineffective. In one study, he looked at 45 of the most widely respected medical studies published during the years 1990 through 2003 that claimed to have demonstrated effective treatments for various ailments. In only 34 cases were attempts made to replicate the original test results with larger samples. The initial results were confirmed in 20 of these 34 cases (59 percent). For seven treatments, the benefits were much smaller than initially estimated; for the other seven treatments, there were no benefits at all. The odds are surely worse for the thousands of studies published in lesser journals. Ioannidis estimates that the majority of published medical research is flawed one way or another.

In one of the most widely read technical papers of all time, "Why Most Published Research Findings Are False," Ioannidis provided a list of warning signs when evaluating a study:

> The sample sizes are small.
> The estimated effects are small.

There were many unreported statistical tests.

There was considerable flexibility in the variables and analysis.

There was a financial incentive or evident agenda.

It is a hot field with researchers racing for publications.

Professor Ioannidis also advises researchers:

Don't go solo.

Don't cherry-pick data.

Don't be fooled by post-hoc reasoning.

A p-value less than 0.05 is usually not enough.

Be transparent. Share your data and methods.

Anyone doubting the importance of scientific rigor should listen to Ioannidis. He has a vision in which scientific progress is not made by steps forward and back, results reported and retracted. Scientific research can be a collaborative and open enterprise, where data are shared, conflicts of interest are contained, and research is only disseminated after quality peer review. He also envisions a world in which scientists are better trained how to do good science. We hope our book helps move the world one step closer towards Ioannidis's vision.

nOCD

Growing up in a middle-income suburb of Chicago, Stephen Smith was a four-sport athlete, playing football, baseball, boxing, and basketball in high school. Football was his favorite, but after injuring two discs in his lower back his senior year, he was no longer recruited by the big football schools. He was smart and hardworking and attended Trinity University, a highly rated liberal arts college in San Antonio, Texas. During his first two years, Smith quarterbacked the football team and got straight A's. Yet, on the inside, he struggled mightily—battling severe obsessive-compulsive disorder (OCD), a chronic neurological condition affecting about 2.5 percent of the global population. He didn't know that he had OCD, but he knew something was terribly wrong.

Obsessive-compulsive disorder is characterized by a nonstop cycle of distressful thoughts and images. To alleviate the anxiety, people with OCD often perform specific actions, called compulsions, that bring temporary relief, but reinforce the affliction. Common compulsions include checking and rechecking whether a door is locked, mentally reviewing past events, and washing hands. Smith spent nine to ten hours a day in his room Googling his symptoms, hoping to find answers that would make his obsessions go away.

He withdrew from school, disassociated from his friends, and became housebound. His life was spiraling out of control. Smith's primary-care doctor referred him to a psychologist, who misdiagnosed Smith's case as "general anxiety" and referred him to a social worker. The social worker told Smith to snap a rubber band on his wrist every time he felt distress, which is ineffective for someone with OCD. When the rubber-band snapping didn't work, she referred him to a Freudian psychologist who told him to move back to Texas. That was also useless advice.

Smith stumbled onto an online forum of others who were experiencing identical struggles and using a clinically proven therapy called Exposure Response Prevention (ERP). Smith was excited. He no longer felt alone, and he now knew that there were evidence-based treatments that worked.

Smith called a local OCD and ERP specialist, but found he would have to pay $350 per session. It was a kick in the gut. He had finally found a remedy, but he couldn't afford it. Smith asked a wealthy relative to cover his specialist expenses for eight weeks and she agreed to help.

After the first session, Smith understood the debilitating chaos and had a plan to get back on his feet, but then he realized that his clinician was never available for help outside the office, which was when his bouts of OCD occurred.

Smith decided to create his own OCD solution: a mobile platform that would allow people with OCD anywhere in the world to get always-on, personalized care and to connect to a support community of others with OCD. Even though he had little money and even less business experience, he worked with thousands of clinicians to build the prototype, which he called nOCD ("no-cd"). He also started fundraising by pitching the product to hundreds of investors, and raised nearly $100,000.

He then transferred to Pomona College in California to learn the skills he needed to make his project succeed: data science and leadership.

Smith assembled a development team and an advisory board of data scientists and OCD clinicians. Unlike Theranos, nOCD is based on procedures that have been carefully tested, with the results published in peer-reviewed journals. In about eight months, they had a fully functional "beta" model that would deliver personalized evidence-based guidance to individuals, based on episode-related data. In addition, the nOCD team developed a suite of ERP exercises and a scheduler that tells its members when to start, stop, change, or even redo an exercise.

Smith and his team released the platform to 705 people with OCD in order to gather user-experience and event data, which are digital footprints collected when users interact with the app. The nOCD data scientists found that the initial participants seldom used the app's clinical functionality during OCD episodes. Follow-up surveys revealed that the participants had a challenging time navigating the platform.

The nOCD development team revised the interface and launched the redesigned app. An event-data comparison showed that users logged significantly more events and, now, survey data showed that users felt that the app was helping them manage their OCD.

With the improved interface, nOCD gained traction quickly. Several mental health clinicians referred nOCD to their patients, and Smith secured a second round of funding. Smith's long-time friend, Gagan Bhambra, not only invested, but also agreed to help Smith grow the service for a year. A week before the start of Smith's junior year at Pomona, nOCD became a live service.

Without a marketing budget, Smith and Bhambra had to identify cost-effective ways to market nOCD. They focused on using social media to build an OCD awareness movement that would generate enough buzz to encourage people to download the nOCD app. Smith and Bhambra also sent a semi-weekly email newsletter to two different audiences: clinicians and nOCD patients. Inside each email were *deep links* for downloading the nOCD app. These deep links showed where the download originated, allowing Smith and Bhambra to gauge the effectiveness of Facebook, Instagram, Twitter, Reddit, the nOCD website, and their newsletter.

Smith and Bhambra applied some sophisticated statistics that Smith was learning at Pomona to data on over 60 different social media variables collected over a 108-day period in 2017. They used half the data to try different model specifications and then used the other half to test their model.

Smith and Bhambra were able to avoid many data science pitfalls, but they did make a few mistakes. For example, they initially collected data once a week, which made the day of the week a confounding factor. They had to start over and collect new data from scratch. But if it weren't for the lessons learned from their early mistakes, they wouldn't have been able to create the model that led them to success.

After class Friday, Smith would often drive his 2003 Toyota Sequoia to hospitals and medical conferences to publicize nOCD. During one cash squeeze, Smith drove 400 miles to San Francisco to sneak into a mental health conference. He found a customer—one customer, but it was enough to keep the servers running—which kept the company running.

One night, shortly after Smith moved into his Pomona dorm room to start his senior year, the team unexpectedly received a surge of traffic at 3 a.m. while they were sleeping. When they woke up, the servers were down, because thousands of people from the United Kingdom had attempted to join nOCD and crashed the servers. How did this happen? An OCD specialist from the Britain went on a show called *This Morning*, the British equivalent of *Good Morning America*, and gave a persuasive testimonial about nOCD's efficacy in helping her practice. Since the nOCD team had a premium server package, they were able to get immediate customer support and get their servers back up the same day. They now knew firsthand that cash is critical for dealing with the unexpected problems that come with growth.

Smith closed his first major contract with a company that wanted to use the nOCD data to study the effectiveness of different OCD treatments. While Smith was finishing his last semester at Pomona, nOCD secured $1 million in financing, which allowed the nOCD team to build and market an even more sophisticated product.

Since then, nOCD has grown rapidly, and is now the world's largest OCD treatment community, helping people in over 100 countries. They've also revolutionized the way research is conducted on OCD. While patients may be unwilling to speak candidly with therapists and often have imperfect memories of the details of their OCD episodes, the data recorded on the nOCD app is likely to be more accurate and complete. Researchers at UCLA, the University of Illinois, and elsewhere are now using the app's anonymous real-time dataset to better understand OCD symptoms and the effectiveness of various exercises and treatments.

Most importantly, they are really helping people with OCD, as evidenced by a solid rating by Psyberguide, the leading nonprofit mental health review. The motto for nOCD is "feel better" and Smith is among those who are feeling better.

Only one move

Jay was just learning to play poker when he saw a televised episode of the *World Poker Tour*. It had a "Poker Corner" segment that sparked Jay's most ambitious data project to date: find a simple and profitable strategy for playing online poker.

At the beginning of each hand of Texas Hold 'Em, the player sitting directly to the left of the dealer puts a small blind of, say, $1 into the pot, and the player two seats to the left of the dealer puts in a big blind that is twice the size of the small blind ($2 in our example). Each player at the table is then dealt two "hole cards" that only they see.

Starting to the left of the big blind, each player takes turns deciding whether to "call" (match the big blind), "raise" (increase the size of the bet), or "fold" (surrender the hand). If the bet is raised, the other players must match the new bet, re-raise, or fold. The bets go clockwise around the table until the highest bet is called by all players who want to remain in the hand, or all but one person has folded.

If more than one player is still in, three community cards ("the flop") are dealt, which are visible to everyone and can be used by each player to combine with their two hole cards to build a five-card poker hand. Another round of betting occurs, starting with the person to the left of the dealer. After this round of betting, a fourth community card ("the turn") is dealt, and there is another round of betting. Finally, the fifth community card ("the river") is dealt, and there is a final round of betting. For everyone who makes it to the end of the hand, there is a "showdown" and the player with the best five-card hand, created from their two hole cards and the five community cards, wins the pot.

The aspect of poker that appealed the most to Jay was that, in the long run, the poker chips flow away from the clowns and towards those who make the best decisions. The poker table is a laboratory, a place where ideas about strategy are tested and the flow of money represents the impartial judgment of the poker gods. Inferior players can win for a while, but eventually, they will lose money, as surely as the dull-eyed zombies sitting at slot machines throughout Las Vegas.

In a world where people still cling to beliefs that vaccinations cause autism and the earth is flat, there is something deeply satisfying about a domain where beliefs are rewarded or punished, based on how well they match reality. False confidence cannot make a poker player a winner. Only a superior playing style can do that.

A memorable point made by professionals on that fateful Poker Corner segment was that when you're a "short stack," with few chips remaining, the only decision that should be made is whether to go all-in (bet all your remaining chips) or fold. Next to this advice was the sad emoji, ☹, presumably because there is a good chance that players with a short stack will soon lose all their chips. In a tournament, it is certainly better to be a "deep stack," but Jay wondered whether having a short stack is a liability in a cash game where a player who loses all his chips can buy more and keep playing.

If you have $100 in chips and you're playing Bill Gates, who has $1 million in chips, it's not like in the movies where, if Bill bets thousands of dollars, you need to throw the keys to your car into the pot in order to call his bet. Instead, the game is effectively the same as if you both have $100 in chips. If Bill bets $1,000, you can "call all-in" and match $100 of his bet, with the extra $900 returned to him. So, why would a short stack be a disadvantage?

In fact, consider what happens when there's a third player, Jill, who also has $1 million in chips and who calls Bill's $1,000 bet after you do. The rules of poker dictate that the main pot contains your $100 and the matching

$100 bets from the two deep stacks, while the extra $900 Bill and Jill bet go into an $1,800 side pot.

Since you can no longer be forced to fold, you can sit back and relax for the rest of the hand and wait to find out if you win the main pot. The millionaires continue to battle each other for the side pot. If one of the millionaires folds, the remaining millionaire wins the side pot and is your only competitor for the main pot. Your chances of winning the main pot have improved since one player has folded and the one who didn't fold may have been bluffing. Being a "short stack" is starting to sound like it's not such a bad deal after all. It might even be an advantage.

Since Jay was still a poker newbie, he liked the idea that, when playing as a short stack, poker can be boiled down to one all-or-nothing decision. Mastering all aspects of this incredibly complex game seemed out of reach. How could he possibly perfect a strategy that takes into account all relevant information, including the cards, the opponents' betting patterns, the size of his chip stack, his position at the table, how the opponents are playing, and how they perceive him to be playing? Finding a profitable short-stack strategy is much simpler than mastering the complete game.

Still, Jay knew several reasons why a profitable short-stack strategy might be elusive. First, conventional wisdom is that the most important things in poker are (1) your opponents, (2) your position at the table, and (3) your cards. To come up with a successful online strategy, wouldn't Jay have to track every online opponent and analyze all of their playing styles? And how can the strategy be simple if it depends on your position at the table? Furthermore, wouldn't savvy opponents quickly identify any simple strategy Jay used and figure out how to beat it? And did he need to worry about collusion among online opponents that wouldn't happen in face-to-face game?

As any self-respecting data scientist would do, Jay thought to himself, "Let's find out." He mapped out his game plan. One obstacle he had to overcome is that he is quite risk averse when it comes to gambling. Practically the only time he ever played casino games was when a promotion offered to reimburse up to $200 in losses or match up to $200 in wins after two hours of play in return for joining a card club. After confirming that the offer was truly risk free, Jay and a friend decided to join and play video poker. With 15 minutes left in the promotion, Jay increased his bet size to $5 and got a great hand: a royal flush. The casino matched $200 of his $4,000 win, and Jay walked out of the casino with his winnings

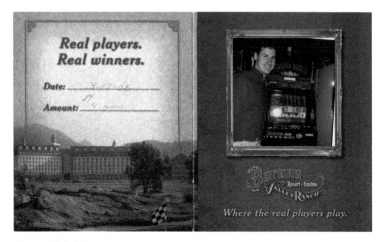

Figure 6.1 Jay's first can't-lose bet

and a picture (Figure 6.1) to prove it, and never returned. Casinos hate people like him.

To test his short-stack strategy, Jay took a deep breath, bought a whopping $50 in poker chips, and started to experiment at tables where the small blind is 25 cents and the big blind is 50 cents. Jay started with a short stack of $5. Since there would be up to nine opponents, his plan was to go all-in whenever he got a hand that ranked among the top 10 percent of all poker hands and fold otherwise. If he lost, he would buy back in for $5 and try again. If he won, he would change tables and start again with $5 (this "hit and run" strategy would be seriously frowned upon in a live game, but online players jump tables so often that it's hardly noticed).

Jay's initial strategy was only slightly profitable, but he knew that this was just the beginning: the data-collection stage. One of the great things about online poker is that you can download "hand history," a file with all the details about the poker games you participated in. These data can then be analyzed to determine how your play could be improved, assuming your opponents continue to play the same way (an untested assumption at this point).

It can be shown mathematically that if you go all-in with a short stack after someone has raised, your opponent is practically forced to call, even if she is bluffing with lousy cards. On average, she will lose more money always folding than always calling, even if she knows her cards are worse than yours at that point. Therefore, Jay knew that he should never bluff in this situation—because a smart opponent will always call his bet.

Based on the hand history he downloaded, Jay determined that the initial hands he should re-raise all-in are ace-queen, ace-king, and pairs of sevens or higher. He was surprised that ace-jack was not profitable, but the numbers were right in front of him. He could expect to lose money if he went all-in with ace-jack after a raise.

After Jay did a similar analysis for cases in which he would be the first raiser, the "System" had been created and was ready to go live. When Jay realized it was simple enough to play at multiple tables simultaneously and that it might be profitable at much higher stakes, he couldn't wait to try it. At dinner, he hesitantly told his wife, "You know how I always roll my eyes when people talk about their 'get rich' schemes…?" She patiently listened to his idea, but was understandably skeptical. After all, isn't everyone trying to make money playing online poker? Why haven't thousands of other people figured out this simple strategy already?

Jay thinks that part of the reason other players hadn't figured out the System is that it is not an optimal strategy. Poker champion Chris Ferguson later told an interviewer that he would not have known how to beat Jay's strategy. He was being modest. In reality, he absolutely would. He could analyze Jay's play and come up with a counter-strategy that would definitely take Jay's chips in the long run.

Jay's strategy is what's called an exploitive strategy, which means that it was engineered to take advantage of inferior play by his opponents. Jay expected astute opponents to adjust (as Ferguson undoubtedly would) and that Jay would have to adjust his strategy in response. So, Jay downloaded hand-history data every month to see if players were changing their strategies to combat the System. They weren't. They played the same way, month after month, and gave money away to the System. By the time the Unlawful Internet Gambling Enforcement Act effectively shut things down, the System had made over $30,000 in profit from Jay's initial $50 investment.

It wasn't always smooth sailing. Poker, by its nature, involves a large amount of luck—so much so, that many people think it is entirely a game of luck played by degenerate gamblers. This perception is bolstered by colorful stories of skillful poker professionals who went broke. However, going broke and being capable of playing the game profitably are not mutually exclusive. Savvy investors and gamblers know about the Kelly criterion, which is a remarkable formula that can give the optimal bet size in order to maximize earnings while minimizing the chances of going broke.

In one study, participants were given $25 and were challenged to make as much money as they could in 30 minutes by making bets on a coin that would land heads 60 percent of the time. Clearly, betting on heads is a winning strategy, but how much should you bet? It turns out that the Kelly formula has an elegant answer: bet the "edge." The edge is the difference between winning and losing chances so the edge for this game is 60–40, or 20 percent. If you bet 20 percent of your bankroll, you can expect to make more money than you would by following any other strategy. Most people in the study bet much more than this, and 28 percent lost all their money, even though the odds were stacked in their favor. Even though the results depended on coin tosses, this was a game of skill in that people who know what they are doing can expect to make more money than those who don't.

Jay estimated what the System's edge would be and he used the Kelly criterion to ensure that he never played at stakes that were too high. This bankroll management ensured that when luck eventually did turn against him, shown in Figure 6.2, his bankroll dropped "only" by $1,800.

During this losing period, Jay worried that his opponents had finally adjusted to his strategy; he analyzed his hand history and found that they

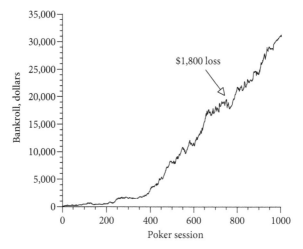

Figure 6.2 Poker has its ups and (occasional) downs

hadn't. It was just bad luck that took a bite out of his bankroll. He resumed playing the System and his luck turned.

Occasionally, Jay's luck was incredible. One time, he was writing emails while playing at multiple tables and was about to click the "send" button, when a poker table jumped to the front of his screen (it does that when it's your turn to act) and Jay accidentally clicked the "raise" button. Unfortunately, he had a bad hand (Jack-8). Fortunately, the next cards dealt gave Jay two pairs and he went all-in, got called by a worse hand, and ended up winning $200. Sometimes it is better to be lucky than good, but it is generally better to be good.

Both Jay and Chris Ferguson provide strong evidence against the claim that poker is just a game of luck. As a challenge, Ferguson once started with $0, played freebie tournaments (it took him several weeks to make his first $2), and then used smart play and careful bankroll management to run his profit up to $10,000. Poker is a game of skill.

Good data scientists know that the starting point is a theory—here, a poker system based on how wagers are made, the rankings of hands, and the logical advantages of a short-stack strategy. After creating his system, Jay tested it against real opponents. That is exactly right, and his system was far more successful than any system derived from coincidental patterns in the data. Instead of considering hands that made logical sense, a data clown might say, "I noticed that if I had gone all-in yesterday with 10–8, 2–2, or Q–7, I would have made more money, so let's add those hands to the list of hands to bet." Jay knew better.

Steve's story

Even though Jay's friend Steve had only played poker for real money once before, he was very interested in learning the System and was willing to invest up to $1,000. Jay taught him the strategy and watched him play for about 30 minutes to make sure he had learned it. He instructed Steve to play at the $1/$2 tables, so that his all-in bet would be no larger than $20, which would allow him to slowly build up his bankroll before raising the stakes. Steve went home, and after an hour, Jay got a phone call.

"I'm taking you guys out to a steak dinner. I'm up $1000!"
"That's not possible, you're only betting $20 at a time."
"I'm looking at my balance right now and it says I'm up a grand."

"Go back to the last table you were playing at."

"Hmm, that's strange, it won't let me buy back in for less than $940."

"The maximum buy-in is supposed to be $200!"

Jay then realized what had happened. Steve was accidentally playing at the $10/$20 tables and was betting $200 per hand instead of $20. It wasn't clear how this could have happened, but Steve's attention deficit disorder may have played a part. To his credit, Steve did play the System correctly and he happened to get lucky. Fortunately for Steve, he never made that mistake again and ended up making over $10,000.

Frank's story

Another friend, Frank, also wanted to learn the System, but Jay was reluctant to share it with him. Frank had a problem. His history of losing money playing online poker had gotten to the point where he actually asked his bank to stop allowing him to transfer money to online sites. Putting him back out on the virtual tables seemed like a clearly bad idea, but Jay started to wonder. Could someone with a gambling problem be "cured" by a winning strategy? As any self-respecting data scientist would do, Jay thought, "Let's find out."

At first, it was looking good for Frank. He made $12,000 in the first month, playing constantly. There were signs that he might crack, like when he lost $4,000 one night playing heads-up against someone who was taunting him in the chat box, but his bankroll was healthy and he seemed to be well on his way to endless profitability. The problem that led to his downfall was boredom.

Some gamblers lose interest—even if they are winning—if there isn't sufficient risk to give them chills and thrills. Following the System and slowly earning money wasn't fun. Frank wanted to play for higher stakes. The highest stakes Jay ever played were the $10/$20 tables, which means that he bet about $200 when he was dealt a hand he had been waiting for. Frank wanted to play at the next level up: the $25/$50 tables, where he would bet $500 per hand.

Jay didn't recommend it, because he had never tested the System at that level, and the higher the stakes, the better the players. Frank might find himself at a table of Fergusons ready to eat his chips. Well, Frank came, Frank saw, and they kicked his butt. That first night knocked him back

$4,500. Jay asked for the hand history and was disturbed to find that, based on how Frank's opponents were playing, no simple hit-and-run strategy would work at the $25/$50 tables. That could be why strategies like Jay's System weren't widely used. The players who were clever enough to figure out the System played at tables where it wouldn't work. The next time Frank and Jay spoke, Frank said he had lost the entire $12,000.

Don's story

Don was also interested in learning the System, even though he didn't play poker and didn't like gambling. He told Jay "as an upstanding actuary, I have absolutely no creativity." Jay told him "you're perfect." Don was the anti-Frank. He was never bored and he was never down more than $2.50. He was perfectly content to play at the lowest stakes. So much so, it started to annoy Don's wife, who joined Jay in pleading for him to move to the higher tables. She didn't want her husband playing games on the computer for $4 an hour. He was eventually shamed into playing for higher stakes.

Don's story has a happier ending than Frank's. He followed the System and made over $2,400 because his actuarial spirit kept his emotions in check.

Unsearched pennies

Long before Jay played poker or even knew what data science was, he collected pennies. The wheat penny is a U.S. one-cent bronze coin minted between 1909 to 1956, with Abraham Lincoln's profile on one side and two stalks of wheat on the back. Most are ordinary, but a few are worth hundreds of dollars. Jay saw that a company was selling "500 unsearched wheat pennies." Jay took that to mean that the company claimed it was selling 500 bronze lottery tickets; even the rarest and most valuable wheat pennies might be in there.

Jay was skeptical and thought of a way to test the advertising claim. He bought a few bags of 500 pennies and looked up the value of each coin, based on the date and the mint where it was produced (Philadelphia, San Francisco, or Denver). His prediction was confirmed: these bags were significantly less valuable than if they were random draws of 500 wheat pennies from the population of pennies produced. The pennies had evidently been searched before being sold. His use of statistics was basic, but there was a budding data scientist deep inside him trying to get out. The need to find scientifically valid answers to questions would infect him and eventually lead to a strong desire to help write a book in the hopes that it would infect everyone else.

Notice, again, that Jay started with a reasonable theory—that the company knew their bags of pennies didn't contain valuable pennies—and he then collected data to test the theory, rather than committing a Texas Sharpshooter Fallacy by sorting the coins into dozens of categories based on the year and mint and then noticing a peculiarity, such as 10 percent more 1942 Philadelphia pennies than expected.

The world has questions far more important than whether or not bags of pennies are a rip-off. Important questions deserve data-based answers. Not the kind of data-driven answers where people torture data to find something that backs up their opinions, but conclusions that are reached after making a testable prediction and then using reliable data to see if it's correct. Good data scientists risk being wrong.

There's a well-known phenomenon in psychology called *hindsight bias*, or the "I-knew-it-all-along" phenomenon. The easiest way to demonstrate the bias is to give 50 people a study that confirms the belief that "absence makes the heart grow fonder," and give 50 other people a study that supports the belief "out of sight, out of mind." Both groups will say that they expected the results they were shown, even though the studies contradict each other.

It is important to form an opinion before looking at data. It is too easy to data mine and p-hack and then say the results make sense, even though you had no idea what to expect. As physicist Neils Bohr said, "Prediction is difficult, especially about the future." It is easy (and useless) to predict things that have already occurred. Good data scientists don't fool themselves into committing a Texas Sharpshooter Fallacy.

The Octagon

In 1993, a high profile (and bloody) experiment was conducted. Eight men with eight different fighting styles entered a single-elimination tournament that would determine the Ultimate Fighting Champion. The fights took place in a fenced-in eight-sided "cage" called the Octagon. The prize was $50,000 and there were only three ways to win: (1) knock out your opponent, (2) force your opponent to "tap out" or submit, or (3) cause your opponent's corner to throw in the towel. There were no time limits, weight classes, or judges. The only rules were no biting or eye gouging. It seemed to be a senseless return to ancient gladiatorial contests, but there was a deeper purpose.

One of the co-creators of the Ultimate Fighting Championship, Rorion (pronounced HO-ree-on) Gracie had come a long way from his days of panhandling in Hawaii and working in fast food restaurants in Southern California. Rorion's father Helio and uncle Carlos in Brazil had modified the martial art taught to them a generation ago by a Japanese judoka named Mitsuyo Maeda (who happened to be a specialist in the wrestling aspects of judo). They turned it into their own style, now called Brazilian Jiu-Jitsu, which focuses on wrestling and submission skills rather than the highlight-reel throws and hurried groundwork that judo is known for. (When Maeda left Japan, his style was referred to as "Kano Jiu-Jitsu," which led to the name Brazilian Jiu-Jitsu, even though it is more closely related to judo.)

The Gracie family had created a new self-defense system which was extremely effective in real fights, but mostly unknown outside of Brazil. Rorion moved to the United States and started giving lessons in his garage before opening up an academy and inviting a few of his brothers to help him teach. In addition to positive word-of-mouth reports and a growing collection of videotaped challenge matches against other martial arts, Rorion wanted to showcase his family's style. He contacted a promoter named Art Davie who came up with the idea of a tournament pitting different martial arts against each other. The Ultimate Fighting Championship (UFC) was born.

The fighters represented a wide variety of styles: savate, sumo, kickboxing, American kenpo, Brazilian Jiu-Jitsu, boxing, shootfighting, and tae kwon do. The obvious choice to represent Brazilian Jiu-Jitsu was Rickson (pronounced "Hickson"), who was the most dominant fighter, technically and physically, in the Gracie family. However, Rorion was confident enough in his style to choose Rickson's younger and smaller brother Royce (pronounced "Hoyce") instead. He did not want a victory to be attributed to superior strength or athleticism. If a small man won, it would be clear that his technique was responsible for the victory.

When the event took place, people didn't know what to expect. Without rules or protection, couldn't someone be killed? The first fight was quick and brutal, with the savate fighter kicking his opponent's tooth through the cage and into the crowd. Royce's first opponent was a boxer, who became completely lost and confused when Royce tackled him and pinned him to the ground. The boxer "tapped out" after only two minutes. Royce's semi-finals match was even shorter, as he choked an experienced and physically imposing opponent into submission in under a minute. Royce won the final match (against the savate fighter who had kicked a tooth into the audience) in less than two minutes, again taking the fight to the ground and strangling his opponent.

The event served its purpose: no one could have anticipated that the Ultimate Fighter Champion would be a skinny guy who barely threw a punch, much less that he would defeat three fearsome opponents in less than five minutes.

Action movies and legends about the effectiveness of traditional martial arts had shaped public beliefs for years, but this experimental evidence was hard to ignore. A new respect for grappling-based styles of self-defense took root.

There were still many skeptics, so they repeated the experiment a few months later. Ultimate Fighter Championship 2 invited 16 fighters, and Royce again reached the finals undefeated. Jay still remembers his thoughts before the final match, which pitted Royce against a frighteningly brutal kickboxer who had crushed his three prior opponents. Jay thought, "I think the little Brazilian guy is going to win, but I have no idea how." His instincts were right, as Royce quickly took the fight to the ground and worked his way to a dominant position, sitting on his opponent's chest. As the kickboxer frantically tried to escape, Royce patiently maintained his position and pecked at him with punches. In frustration, the kickboxer tapped out and Royce was again the UFC champion.

By now, the public was impressed by Brazilian jiu-jitsu, but there was still an open question. Could Royce defeat a wrestler? He generally won by taking his opponent to the ground, gaining a dominant position, and then applying a finishing hold, like a choke or arm-lock. It's one thing to control stand-up fighters who are not used to being on the ground, but quite another to do so against a seasoned grappler.

The answer came nine months later when Royce faced a 260-pound Greco–Roman wrestler in the finals of UFC 4. (Royce didn't win UFC 3; after submitting his first opponent, he was too exhausted to compete in the second round.) With an 80-pound weight advantage, it was no surprise that the wrestler took Royce down and maintained the top position. The fight went on and on, for 15 minutes with nothing changed: Royce was on the bottom, taking strikes from the wrestler and offering nothing in return other than holding tight to minimize the damage. Just as the announcers were writing him off, saying he had the "heart of a champion" for not giving up, Royce wrapped his legs around the wrestler's neck in a "triangle choke" and forced the wrestler to tap out. It was a remarkable moment. The world was shown that fighting is like chess—and there is a family of Brazilian grandmasters.

Royce's family started fighting around the world (including Rickson, who despite never losing, disappointed fans by retiring after the tragic loss of his son), and Brazilian Jiu-Jitsu schools started popping up

Figure 6.3 Jay and Royce. Does Royce look like he could be the "ultimate fighter"? Photo by Catherine

everywhere (Jay even took lessons for a few years). Nowadays, it's almost impossible to find a UFC fighter who hasn't trained extensively in it. Commentator Joe Rogan once said that martial arts evolved more in the ten years following UFC 1 than in the preceding 700 years. It was truly a revolution.

Rorion Gracie and Art Davie would probably not consider themselves data scientists, but their scientific mindsets led them to an experimental design that was similar to what a good data scientist would have recommended. The tournament format ensured that unbeaten fighters would have a fair chance to challenge this new martial art. (We would have recommended the slight improvement of having a public drawing of the first round matchups.) The almost complete absence of rules shut down any criticism that the fights were not similar enough to street fights to be considered valid real-world tests.

Video tapes of Gracie fights before the UFC matches could have been misleading if there were losing fights that were not included on the videos—like Derren Brown's unsuccessful coin flips. The live broadcast of the UFC events eliminated speculation that the public was only shown cherry-picked results. The dramatic growth in popularity of Brazilian jiu-jitsu after those first UFC matches demonstrates that live results are much more compelling than videotapes of past events.

What would be the ideal experiment to demonstrate the effectiveness of Brazilian Jiu-Jitsu? One possibility would be to select a large group of novices and teach Brazilian Jiu-Jitsu to a random half for ten years, while the other half were trained in other styles. The two teams would then fight it out and the number of wins and losses would be tallied.

What happened in the real world was fairly close to this experimental design. Being born as a Gracie family member is practically a random assignment unless there's some kind of genetic fight gene that they possess. Based on Joe Rogan's comment that the revolution of martial arts occurred in the 10 years following UFC 1, Table 6.1 shows how the Gracie fighters did during those 10 years:

The chances that a random group of fighters would be this successful are less than one in a million. Since the Gracies are related to each other, their success might be due to confounding factors such as athleticism or fighting spirit. However, the fact that they all trained in a family martial art is a more compelling explanation. The evidence is strong: Brazilian Jiu-Jitsu works.

Table 6.1 *The Gracies are far above average (Sherdog.com)*

	Win	Loss	Draw
Carlson Gracie Jr.	0	0	2
Crosley Gracie	1	0	0
Daniel Gracie	3	1	0
Ralph Gracie	6	0	0
Renzo Gracie	10	5	1
Rickson Gracie	11	0	0
Royce Gracie	12	1	2
Royler Gracie	3	1	1
Ryan Gracie	3	2	0
Total	49	10	6

On Neil DeGrasse Tyson's show, StarTalk, Jiu-Jitsu was called "the most scientific form of martial arts in the sense that it can respond to experimental results." Like true scientists, Art Davie and Rorion Gracie tried to anticipate criticism by creating an experimental design that minimized any perception of unfairness. Royce was one of the smallest fighters and not the most technically advanced in his style. (He once said, "my brother Rickson is ten times better than I am.") The opposing fighters were even allowed to tape their wrists so that they could throw harder punches without injuring themselves (Royce didn't bother since he relied more on submission holds). The Gracies were willing to put their legacy on the line, staging multiple contests against dangerous fighters. It wasn't long before the experimental results were in and observers became believers.

Wishful thinking

A study asked 100 high-school students to predict their scores on a math test. The average predicted score (75) was higher than the average actual score (60), but there was a 0.70 correlation between the predicted and actual scores.

The author drew two conclusions. The first was that students overestimate their ability. We personally believe that positive reinforcement is better than harsh criticism, but this is a clear example of people seeing what they want to see. Instead of overestimating their ability, perhaps these students underestimated the difficulty of the test that they would be given. It would have been more interesting to ask students to estimate their percentile score relative to the other students in the class. If 70 percent of the students think they are above average, this would be convincing evidence that they overestimate their ability.

The author's second conclusion was that test scores can be increased by raising students' self-esteem:

The positive correlation between the predicted and actual scores demonstrates that scores are boosted by self-esteem, so teachers should increase their students' self-esteem. This is confirmed by the fact that the two students who predicted that they would fail did fail. An inferiority complex prevents success.

The positive correlation between the predicted and actual scores does not prove that scores are affected by expectations, but may instead reflect the fact that students know something about their mathematical abilities. Most students who did well surely knew they are good at math. The two students who failed knew that they did not know the material well enough to pass the test. They weren't being unduly pessimistic; they were being realistic. A valid test might divide the students randomly into two groups, one of which is given positive reinforcement, while the other is given none (or criticized for their inadequacies). This study was not a valid test.

LEAPS

Pakistan is the fifth largest country in the world, with a population over 200 million, and a per capita income of around $5,400, placing it among the bottom third. Forty percent of the population over the age of 15 is illiterate.

Nicholas Kristof is a *New York Times* columnist who has been a Pulitzer Prize finalist seven times and won twice. In 2010 he wrote that,

The public education system, in particular, is a catastrophe. I've dropped in on Pakistani schools where the teachers haven't bothered to show up (because they get paid anyway), and where the classrooms have collapsed (leaving students to meet under trees). Girls have been particularly left out. In the tribal areas, female literacy is 3 percent.

Three Pakistani economists, Tahir Andrabi, Jishnu Das, and Asim Ijaz Khwaja, have spent nearly 20 years studying Pakistan's education system, trying to identify practical ways to fix what is clearly broken.

Experiments are impractical in most studies of human behavior. We can't make people marry, divorce, or have children and see how their lives change. Instead, we have to deal with the self-selection bias that can occur when some people choose to marry, divorce, or have children, and others don't.

However, these economists were able to conduct a variety of valid educational experiments in a project named Learning and Educational Achievement in Punjab Schools (LEAPS) funded by the World Bank, the National Science Foundation, and other organizations.

One experiment involved 112 villages in the rural Punjab province in Pakistan, with an average of 7.3 schools per village, some private and some public. Families rarely send their children to schools outside their village, but they know little about the schools in their village other than the cost. Many families assumed, reasonably enough, that more expensive schools are better schools—perhaps high-cost schools spend more money on teachers and supplies, or perhaps they can charge more because there is more demand for the high-quality education they provide.

The economists reasoned that parents would be able to make better-informed choices if they had real evidence of school performance, and that schools might respond to parents making better-informed choices by trying harder to deliver a quality education. So, the economists collected standardized test scores and did an experiment by randomly separating villages into two groups. In half the villages, parents were sent report cards showing their children's performance and the performance of all the private and public schools in the village. The other half was the control group, and parents did not receive report cards.

The results were astonishing. In the villages receiving report cards, many of the worst performing private schools went out of business, and the remaining schools (both private and public) made progress. Test scores increased by 42 percent compared to schools in the control villages, while private school tuition dropped by 17 percent. High-priced schools evidently could not rely on their reputation to attract students. They had to offer a better education, or make it less expensive.

In another set of experiments, villages were randomly separated into three groups. In the first group of villages, a single, randomly selected private school was given an unconditional cash grant of 50,000 rupees (worth about $500 at the time). Private schools in rural Punjab generally have modest budgets, with median annual revenue of only $3,300, so a

$500 grant is a big deal. In the second set of villages, every private school was given 50,000 rupees. No grants were given to schools in the third group of villages.

Even though the grants were unconditional, the schools invested the money in their schools. They evidently wanted to spend more, but were unable to do so because their resources were limited. Schools in the first group (one grant per village) mostly spent their grants on desks, chairs, and computers that enabled them to increase their enrollment and revenue; there were no changes in their test scores. In addition to facilities improvements, schools in the second group (grants to all private schools) increased teacher salaries, evidently to retain and recruit good teachers since the schools were competing against other schools that also received grants. Enrollment increased by less than half that of the first group and, in contrast to the first group, test scores increased.

While a private individual might prefer giving money to a single school, the economists argued that government grants to all schools are socially preferable because these increase teacher quality and test scores.

An educational revolution is happening in Pakistan, supported by evidence-based experiments. A third of all students in Pakistan now go to low-cost private schools which, overall, outperform government schools by all sensible measures. Female literacy is surging and female college students now outnumber males.

It had to win somewhere

The spirit of experimentation and the desire to find objective answers to questions should be in the bloodstream of every data scientist. Otherwise, erroneous conclusions and poor decisions are likely.

A web company TrySomething conducted a randomized experiment to test a new layout of a web page across 100,000 domain names and found that, overall, the original layout was better. The manager decided to use the new layout on every domain where it had outperformed the old layout. Why was this a Pitfall of Data Science?

Data scientists should be concerned that there was no expectation or logical reason why the new layout would do better on those particular domains. It may just have been a lethal combination of luck and p-hacking. The vast majority of the domains had only a few ad-clicks per week, so winning by a margin of 3 clicks to 2 was meaningless. The new layout was bound to have a few hits, since it effectively got a hundred thousand swings of the bat.

There was a lively discussion about why using the new layout on the selected domains wasn't a good idea, but there was no need for debate. The issue could easily be settled by running a new experiment on the domains where the new layout had done well. Every time a web visitor showed up, have the testing platform flip an electronic coin to determine whether to display the original layout or the new layout. If the results from the first experiment were meaningful and not the result of some Texas sharpshooting, then the new layout should continue to be more profitable. Don't argue; verify.

Unfortunately, the company's managers decided that further experimentation wasn't necessary (after all, up is up). They ordered the changes made, and profits probably suffered.

Data scientists should not be satisfied with naval gazing. They should test poker strategies instead of assuming that conventional wisdom is correct. They should find out if their suspicions about misleading ads are justified. They should risk defeat in order to demonstrate that they are right. They should retest when the initial results are unconvincing.

Above all, they should state their theories and expectations beforehand, before crunching numbers. Data scientists live up to the "scientist" moniker if they use experiments to test their theories, instead of being Texas Sharpshooters.

Jay's teenage daughter has already demonstrated a scientific mindset. Jay told her, "data science works! In the two days after I analyzed the strategy of the top players on Castle Crush and adjusted my cards, my rating went from 2,400 to over 3,000 and got me into the top castle for the first time!" She thought for a second and responded, "You don't know for sure; you didn't have a control." Jay nodded in appreciation: "I like the way you think."

Unintentional clowns

Clowns fool themselves. Scientists don't. Often, the easiest way to differentiate a data clown from a data scientist is to track the successes and failures of their predictions. Clowns avoid experimentation out of fear that they're wrong, or wait until after seeing the data before stating what they expected to find. Scientists share their theories, question their assumptions, and seek opportunities to run experiments that will verify or contradict them. Most new theories are not correct and will not be supported by experiments. Scientists are comfortable with that reality and don't try to ram a square peg in a round hole by torturing data or mangling theories. They know that science works, but only if it's done right.

The Sixth Pitfall of Data Science is:

Fooling Your.self

Confusing Correlation with Causation

> "As you can see from the charts, revenue went up 20 percent when we put our keywords on your web pages."
> "That's funny, we ran a randomized experiment during that time and revenue was actually 5.9% lower when your keywords were used."

Two men flew all the way from Australia to the ABTester headquarters in downtown Los Angeles to give an important sales pitch. If their pitch was persuasive, their company would gain a tremendous amount of web traffic. In a large glass-walled conference room, they presented their PowerPoint slides with confidence and pizzazz, including charts displaying the dramatic improvement in revenue that occurred after their proprietary keywords were displayed on many web pages.

There were only two ABTester employees in attendance: a manager and a data scientist. The data scientist was skeptical. Why had these two Aussies flown all the way from Down Under after he had already emailed them evidence that contradicted their claim? They must be good talkers—or at least think they are good talkers.

When the presentation was over, the data scientist presented ABTester's view of the data. ABTester ran an experiment on the same domain names during the same time period displayed in the Aussies' charts. Half the time (at randomly determined times), they used the Australian company's keywords; half the time, they used ABTester's original keywords. The experimental

The 9 Pitfalls of Data Science. Gary Smith and Jay Cordes. Oxford University Press (2019).
© Gary Smith and Jay Cordes 2019. DOI: 10.1093/oso/9780198844396.001.0001

results showed clearly that revenue was 6 percent higher with ABTester's keywords than with the Australian keywords. The Australian company had demonstrated correlation (revenue increased after the start of the experiment), but ABTester had shown causation (revenue increased more when the original keywords were used). There would be no deals struck that day, because causation beats correlation every time.

The lesson didn't seem to stick with the manager, however, because the manager found himself on the other end of the debate a few months later when he and an analyst took credit for an increase in revenue from Chinese web traffic after they switched keywords on several domain names. Up is up.

The same skeptical data scientist tested their claim by finding domain names that had not been changed. When the daily revenue for the two sets were plotted next to each other, the lines were indistinguishable. Revenue had gone up equally everywhere; the keyword changes hadn't mattered at all. Sometimes, up doesn't mean anything.

What causes what?

A 2018 survey of 1,000 Millennials between the ages of 22 and 37 found that Millennials living at home with their parents had less income and savings than did Millennials living on their own. The conclusion:

Contrary to received wisdom, it may be smarter for Millennials to fly the nest sooner than later, if only for the greater wealth of opportunities that could expand their earning power.

An obvious alternative explanation is that it is not that Millennials have fewer job opportunities because they live at home, but that they live at home because they don't have well-paying jobs.

When two things are correlated, it may just be a coincidence—a temporary fluke that will soon vanish. If there is a causal relationship between two things, it could be that the first affects the second, the second affects the first, or some third factor affects both.

The power of time

When any two things increase over time, there can be a statistical correlation without any causal relationship between them. Figure 7.1 shows beer sales and the number of married people in the United States. The correlation is

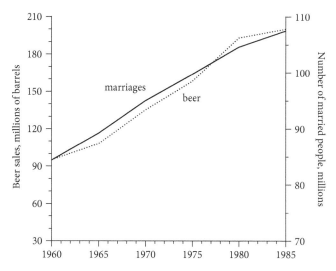

Figure 7.1 Drinking and marriage

a remarkable 0.99. From this strong correlation, can we then conclude that drinking leads to marriage? Or perhaps the other way around, that marriage leads to drinking?

While there are surely a few folks who would agree with one or the other interpretation, the real explanation is almost certainly that as the population has grown over time, so has beer consumption. And so have marriages, babies, automobiles, shoes, college attendance, heart disease, and many other things. Does wearing shoes increase the number of babies? Does driving cars cause heart disease? There isn't necessarily any relationship between things that increase with the population—other than that they increase with the population.

Many things that are otherwise unrelated increase over time because the population, income, and prices are increasing. This simple fact is the basis for many spurious correlations. Stork nests and human births increased over time in northwestern Europe, evidently supporting the fairy tale that storks bring babies. The number of missed days at work and the number of people playing golf increased over time. Do people skip work to play golf or do people get injured playing golf?

The wonderful website spuriouscorrelations.com has many other examples, such as correlation between the number of lawyers in Nevada and the number of people who died after tripping over their own two feet.

Is correlation enough?

Most people do not appreciate the fact that random events (like coin flips) are likely to show striking patterns. This misperception underlies the mistaken belief that a pattern discovered by a computer must be meaningful, when it may be meaningless. This is an inescapable problem with data mining.

Humans have the capacity to be skeptical of data that support fanciful claims. Computers do not, because they do not have the wisdom or common sense needed to distinguish between the plausible and the far-fetched. A computer would not question a statistical relationship that had been discovered between Nevada lawyers and death by clumsy walking.

An article by Cristian S. Calude and Giuseppe Longo titled, "The Deluge of Spurious Correlations in Big Data," proves that, "Every large set of numbers, points or objects necessarily contains a highly regular pattern." Not only that, but the more data,

the more arbitrary, meaningless and useless (for future action) correlations will be found in them. Thus, paradoxically, the more information we have, the more difficult is to extract meaning from it. Too much information tends to behave like very little information. If there is a fixed set of true statistical relationships that is useful for making predictions, the data deluge necessarily increases the ratio of meaningless statistical relationships to true relationships.

We think that patterns are unusual and therefore meaningful. In big data, patterns are inevitable and therefore meaningless.

Sometimes it is not possible to design and conduct an experiment the way ABTester did to test for a causal relationship, and the associations in the data are all we have. Then humans must use theories, wisdom, common sense, and critical thinking to assess whether a statistical correlation is real and useful, or coincidental, temporary, and useless.

In the 2008 *Wired* magazine article mentioned in Chapter 3, Chris Anderson argued that we no longer need to think about whether statistical patterns make sense: "Google conquered the advertising world with nothing more than applied mathematics. It didn't pretend to know anything about the culture and conventions of advertising—it just assumed that better data, with better analytic tools, would win the day." He went on:

There is now a better way. Petabytes allow us to say: "Correlation is enough." We can stop looking for models. We can analyze the data without hypotheses about what it might show. We can throw the numbers into the biggest computing clusters the world has ever seen and let statistical algorithms find patterns where science cannot.

We doubt that Google's empire was built on correlations, unless he meant correlation in the sense that ABTester's experiments tested whether certain keywords were "correlated" with higher revenue. ABTester didn't have a theoretical model for *why* certain keywords performed better, but the keywords were not data mined by being chosen after the experiment. ABTester conducted a scientifically valid experiment to establish causality for words selected before the experiment. There's an important distinction between discovering statistical correlations and demonstrating causation with a randomized controlled experiment.

By randomizing the use of the two pre-selected sets of keywords, ABTester's experiment neutralized confounding influences and demonstrated that the observed differences were real. That is the scientific method, and it's not obsolete, especially at Google. In 2009, the year after the "End of Theory" article was published, Google reported that it ran about 12,000 randomized experiments. Google did not become a preeminent tech company by basing its decisions on coincidental correlations.

Judea Pearl, 2011 winner of the ACM Turing Award, the "Nobel Prize of computing," argues that for machines to be truly intelligent, they need to understand cause and effect:

As much as I look into what's being done with deep learning, I see they're all stuck there on the level of associations. Curve fitting. That sounds like sacrilege, to say that all the impressive achievements of deep learning amount to just fitting a curve to data. From the point of view of the mathematical hierarchy, no matter how skillfully you manipulate the data and what you read into the data when you manipulate it, it's still a curve-fitting exercise, albeit complex and nontrivial.

When will computers move beyond curve fitting to understanding? To developing theories? To planning and conducting experiments? We don't know, but until they do, computer algorithms need good data scientists as much as data scientists need computers.

Tanks and clouds

When an enemy hides camouflaged tanks in a forest, there may not be any soldiers watching the forest and, if there are, they may not notice the tanks. The U.S. Army wanted to install surveillance cameras equipped with an advanced deep neural network algorithm that would recognize tanks hiding among trees. Researchers took 200 photographs of a forest, 100 with tanks and 100 without. The algorithm was then trained by analyzing 50 photos with tanks in the forest and 50 photos with no tanks. After the training, the algorithm was shown the remaining 100 photos in order to validate the algorithm's tank-identification capability. It was 100 percent accurate.

The researchers were shocked when the Pentagon retested the algorithm with new photos and found it was no better than a coin flip. It turned out that the original photographs of an empty forest were all taken on a sunny day, while the photographs with tanks in the forest were taken on a cloudy day. The algorithm had no way of knowing that it was supposed to identify tanks, and the biggest difference between the pixel patterns in the "tank" and "no-tank" photos was the presence or absence of clouds. The tank detector ended up being a cloud detector.

The point is not that computers cannot be trained to put "tank" labels on tank photos most of the time, it's that computer algorithms sometimes fail miserably because labeling pixel patterns is very different from humans understanding what things are. Focusing on persistent, but meaningless, differences is a common pitfall of a blank-slate approach, in which computers are given data and try to identify patterns on their own without human assistance; that's why the tank algorithm focused on clouds. A better way to test the algorithm would be to make copies of the tank photos with the tanks edited out. That way, if the algorithm categorizes the photos correctly, we can be certain that it was because of the presence or absence of tanks. Data scientists who expect computers to categorize pictures as tanks or no-tanks, without any guidance, have their heads in the clouds.

The Retinator

The leading cause of blindness for working-age adults is diabetic retinopathy (DR), a medical condition that damages the retinas of diabetes patients. Due to the increasing number of people with diabetes, there were 800,000 cases of DR-caused blindness worldwide in 2010, a 27 percent increase since 1990.

Diabetic retinopathy has no apparent symptoms, so it is important for people with diabetes to have their eyes checked regularly by an ophthalmologist or optometrist who looks for tiny hemorrhages and other abnormal features in the retina. Fewer than half of all people with diabetes do this, many simply because it is a hassle to get these exams.

Michael Abramoff, MD, PhD, is a retinal specialist who legally emigrated from the Netherlands to the state of Iowa in the United States. He envisioned a potential solution to this problem: an autonomous artificial intelligence system that would be capable of diagnosing a patient with DR without the need for a specialist. Not only might this reduce the number of incidents of unnecessary blindness, it might reduce healthcare costs while improving the quality of healthcare.

Dr. Abramoff's journey in developing this technology and getting it to patients was a long one. He told us that he started working with neural networks 30 years ago, and he was drawn to the idea of learning lessons from biology instead of relying on computer curve fitting to build an autonomous AI system: "Evolution has led to a refined system, and it works."

Twenty years ago, Dr. Abramoff demonstrated that computers could be used for early detection of DR and he thought he would have the technology perfected in a few years. However, three obstacles slowed his progress. First, cheap sensors to produce digital imaging were not widespread yet; in retrospect, he considers this to have been the most important constraint on the adoption of autonomous AI in healthcare. Second, computers were not yet capable of analyzing the necessary data in anything resembling real time. Third, it was not always possible to determine the "clinical truth" about the presence of DR that is needed to check a computer algorithm's diagnosis. Even today, there are limits to what can be measured objectively, and doctor diagnoses are sometimes incorrect.

After decades of groundbreaking research, hundreds of research articles, and over a dozen patents, Dr. Abramoff became known as a world authority and earned the nickname "the Retinator." His approach is to classify images of the eye the way retinal experts do, by looking for specific signs of DR. He developed multiple detectors to look for known predictive features of DR such as hemorrhages and other biomarkers. He also wanted his results to be comprehensible so that doctors and patients understand the computer diagnosis.

He told us that if his system failed to recognize a case of DR, he wanted to know why it failed. It was also important to him that the pathological mechanisms for predictive features were known. He wouldn't settle for a black box, because he knew that risked catastrophic failure. Dr. Abramoff and his team went through the long process of documenting every line of code and identifying every risk before making his program available to the public. He founded a company named IDx which produces the diagnostic system called IDx-DR which classifies DR about as well as a specialist. He told the *Review of Ophthalmology* that, "If I give clinicians an image with a bunch of hemorrhages, they'll say, 'This is likely diabetic retinopathy.' If I start taking those hemorrhages away, eventually they'll say, 'There's no disease here.' Biomarker AI systems like ours work similarly."

Dr. Abramoff wasn't the only one to see the transformative potential for a computer diagnosis of DR. Others were more attracted to a black-box algorithm, and tackled the problem with a bottom-up, blank-slate approach, letting computers teach themselves what characteristics are useful for diagnosing DR. In theory, this approach might find important features that clinicians did not know were significant. Ursula Schmidt-Erfurth, MD, a professor from Austria, described the knowledge-discovery approach this way: "it's a different strategy to let an algorithm search for any kind of anomaly from a large sample of normal and diseased cases.... The advantage of this approach is that it's an unbiased search and can find a lot of relevant morphological features. Of course, one then has to correlate these previously unknown features with function or prognosis to make sense out of them."

However, these kinds of data are very complex and much of this complexity is irrelevant, so many features found to be correlated statistically with a DR diagnosis may be spurious. As with the tank detector that was a cloud detector, systematic errors can arise when algorithms consider nothing but unfiltered images. Another example involved an AI program that was trained to tell the difference between malignant and benign tumors by examining pictures of both. The malignant tumors had rulers next to them that doctors had used to measure the size of the malignant tumors and, not knowing what they were looking for, the AI program focused on the rulers rather than the tumors.

Here, for a black-box DR diagnosis, there might be a bias due to the color of the retina, a different kind of background, or even part of the border around the image. A black-box model can fail with new images, with no one knowing why it failed.

In April 2018, the FDA approved the first autonomous AI diagnosis system: it was Dr. Abramoff's IDx-DR system.

We are hopeful that the Retinator's approach paves the way forward in other areas. We would be more comfortable sitting in a self-driving car that recognizes the known shapes and locations of stop signs, instead of a bottom-up algorithm that relies on pixel matching and might be derailed by an unexpected peace sticker. (Note: we would rather be sharing the road with either kind of self-driving car than with human drivers who have alcohol-impaired, sleep-deprived, phone-obsessed reaction times, but that's a pretty low bar).

Poker tendencies

In 2009, a mysterious poker star appeared in the online poker universe. Someone with the username "Isildur1" won hundreds of thousands of dollars playing poker against all-comers. This newcomer drew the attention of a famous high-stakes player named Tom Dwan, and soon these two were playing head-to-head at six simultaneous tables at ultrahigh stakes. Isildur1 won $3 million dollars in four days, and the poker community was stunned.

Isildur1 continued to play the highest stakes against the biggest names in poker and was involved in all ten of the largest online poker pots ever played (the biggest hand was for $1.3 million) and his total profit grew to $6 million. Who was this mystery player who was taking down the best poker players in the world? Then, suddenly, all the profits evaporated, with the final blow being a $3.2 million loss during one crushing evening against Brian Hastings. Soon after this defeat, Isildur1 was identified as a 19-year old Swedish player named Viktor Blom. After losing all his winnings, he disappeared from the tables for a while. However, the story of Blom's defeat continued to unfold.

Hastings later revealed that he and two other players (Brian Townsend and Cole South) had prepared for his showdown match with Blom by studying how he had played 30,000 hands. Hastings didn't reveal what they were looking for or what they found, but it was evidently useful. This teamwork made headlines because, technically, it violated the terms and conditions of the poker site.

Aside from the controversy, this is a very nice example of the difference between educated data analysis and blind data mining. Hastings, Townsend, and South had expert knowledge of the successful poker strategies used by experienced professionals, and they no doubt looked for exploitable deviations in Blom's play.

Poker is all about making good decisions. Suppose you have a pretty good hand and an opponent places a large bet. Do you call, raise, or fold? It is not just a question of how good your hand is, but an assessment of your opponent's hand. Did he make a large bet because he has a good hand, or is he trying to bluff you into folding? You should think about the way he has bet his hand so far, and also about his tendencies; for example, how often does he bluff, and when?

Hastings, Townsend, and South didn't ransack their data, looking for anomalies. Maybe Blom happened to be dealt a pair of Queens somewhat more often than expected and happened to be dealt a pair of Kings less often than expected—suggesting that when he bets aggressively he is more likely to have a pair of Queens than a pair of Kings. A data-mining algorithm might discover such an anomaly and consider it important. Expert poker players wouldn't notice something like that and, if they happened to notice it, they would know it doesn't matter.

Hasting's team knew in advance that poker players must decide when to bet a hand and when to fold and it makes sense that some players are more aggressive than others. They might look at Blom's hand history in order to calibrate how aggressive he was. This is the traditional use of statistics: begin with a theory and then use data to test the theory or to quantify a relationship.

It is legitimate to test or estimate things that make sense. For example, it makes sense that a certain baseball player may be less successful hitting certain types of pitches (like curve balls) thrown in certain locations (like low and inside). A valid statistical analysis would quantify these tendencies—and these tendencies would be useful for deciding what pitches to throw to this batter. Worthless data mining might be an unfettered search for patterns that discovers a worthless anomaly like this player happened to hit an unusual number of home runs in Tuesday games.

Similarly, it makes sense that a certain basketball player may prefer to dribble to the right and is more likely to make shots taken from certain spots on the floor. A valid statistical analysis would quantify these tendencies—and these tendencies would be useful for deciding how to defend this player. Worthless data mining might be an unguided ransacking of the data that discovers that this player happened to make more of his shots when his team's score was an odd number instead of an even number.

Another example is penalty kicks in soccer. A person taking a penalty kick must decide where to aim the kick, most commonly to the upper or lower, left or right, corner of the net, as in Figure 7.2—though players occasionally aim for the center.

Players do not want to be completely predictable by always aiming for the same location, but different players surely have different preferences that are revealed in their tendencies. A valid statistical analysis would see if a certain player has tendencies and estimate those tendencies, as in Figure 7.3.

Teams might do more complicated analyses looking for tendencies against certain goalies or in certain situations (such as tie games). These are sensible and potentially useful statistical analyses.

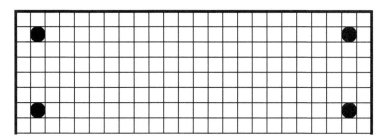

Figure 7.2 Good penalty kick targets

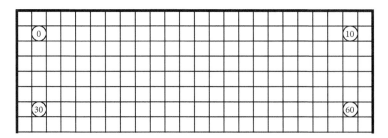

Figure 7.3 Penalty kick tendencies

A data-mining knowledge discovery approach might stumble upon an odd, and useless, pattern, like a certain player happened to have aimed for the lower right corner more often in Saturday games than in Sunday games. You may say to yourself, "of course I wouldn't do any of this silly knowledge discovery involving days of the week or odd or even numbered scores, because that's clearly ridiculous." It's only ridiculous because you know that these variables don't make sense. The danger lurks when you're analyzing variables you don't understand. It's easy to find correlations and it's easy to imagine that they're meaningful. And, if a black-box computer algorithm does the knowledge discovery, there is no possibility of determining whether the variables make sense.

The important distinction in all these examples is that we know ahead of time what we are looking for, in contrast to knowledge discovery, where we are hoping that the data will reveal unexpected patterns. Experienced poker players, baseball players, basketball players, and soccer players know what tendencies to look for—because people differ in these tendencies and because these variations can be exploited. Data miners do not know what they are looking for; indeed, they say that knowledge discovery would be hampered if it restricted itself to logical patterns.

Making peace with losses

Classical economic theory assumes that people make dispassionate, rational decisions. Behavioral economists, in contrast, acknowledge that human decisions are influenced by a variety of psychological and emotional factors. One example of the difference between behavioral and classical economic theory relates to how people react to losing money. Classical economic theory says that we should ignore our losses, which are in the past and cannot be undone. Behavioral economic theory says that people who have lost money may try to "break even" by winning back what they have lost. Thus, Daniel Kahneman and Amos Tversky, two behavioral economics pioneers, argued that a "person who has not made peace with his losses is likely to accept gambles that would be unacceptable to him otherwise." As evidence, they observe that at race tracks, betting on long shots tends to increase at the end of the day because people are looking for a cheap way to win back what they lost earlier in the day.

Gary and two of his students tested this break-even theory by studying how experienced poker players react to large losses. Their data came from

Full Tilt Poker, an online poker room that was reported to be the favorite site of professional poker players. The students were writing their senior theses during the spring 2008 semester, so they looked at data from January of 2008, through May of 2008. They looked at no-limit $25/$50 blind tables with a maximum of six players at a table, and considered a hand where a player won or lost $1,000 to be a significant win or loss. Their data included 346 players. The median number of hands played by these players was 1,738. Half of the players won or lost more than $200,000, ten percent won or lost more than $1 million.

They wanted a numerical measure of each person's style of play so that they could see if people change their playing style after a big win or loss. One generally accepted measure of a player's style is *looseness*: the percentage of hands in which a player voluntarily puts money into the pot. At the six-player tables used in this study, people are typically considered to be very tight players if their looseness is below 20 percent and to be extremely loose players if their looseness is above 50 percent. The average looseness for the players in this study was 26 percent, which is a reasonable number since professional poker players have a reputation for relatively tight play.

Gary and his students found that two-thirds of the players played looser after a big loss than after a big win, confirming Kayhnemann and Tversky's theory.

Is riskier play punished? It seems likely that experienced players generally use sound strategies, and that any changes they make are a mistake. That's what happened. Players who played looser after a big loss were less successful than they were with their normal playing style.

This study was good data science. They had a theory and used data to test and quantify the theory. It's also potentially useful for poker players to know how people react to losses and how they themselves react.

The observed changes in poker play (and their unfortunate consequences) may be applicable to investment decisions. If investors are like poker players, their behavior might well be affected by large gains and losses, for example, making otherwise imprudent long-shot investments with the hope of offsetting prior losses cheaply. Informed by good data science, investors might recognize the lure of break-even behavior and resist the temptation.

A February 2009 article in the *Wall Street Journal* reported that many investors were, in fact, responding to their stock market losses by making increasingly risky investments:

the financial equivalent of a 'Hail Mary pass'—the desperate attempt, far from the goal line and late in a losing game, to fling the football as hard and as high as you can, hoping it will somehow come down for a score and wipe out your deficit.

Data scientists can warn against Hail Mary investments.

Bargaining power

One of Gary's students (Lee) wrote her senior thesis on how household spending decisions in the aftermath of the 2008 stock market crash depended on a husband and wife's relative bargaining power, as gauged by the share of household income earned by each spouse. To simplify, we will just divide households here into male earner and dual earner. This was a path-breaking study in that traditional household spending models do not consider the relative bargaining power of husbands and wives. Lee hoped to have her senior thesis published in a peer-reviewed journal, the first of many during her career.

Gary asked Lee what she expected to find. She said that maybe dual-earner households would spend more or less money on furniture. Gary asked which it was—more or less? Did she expect furniture spending to go up or down? Lee said that she "had no idea." The data would tell her. So it was with every possible household response to the stock market tumble. Lee planned to look at the household data and see which categories varied the most across households with different relative bargaining positions.

The problem is that, with hundreds of categories, there were bound to be some coincidental differences. A data clown might report, "I found that, in dual-earner households, spending on movies went up while spending on new clothing went down." The data are what the data are. Up is up. Or the data clown might invent an explanation after the fact: dual-earner households were more likely to save money on clothing and then go to movies to forget about their troubles. With no compelling theory—just made-up explanations or no explanation at all—a data-mined model would be useless.

Gary persuaded Lee to investigate theories of dual-earner behavior before she looked at her data. She found that there were, in fact, some persuasive theories—not about spending on furniture, clothing, and movies—but about overall spending. She found that women, on average, feel greater vulnerability than men because they earn, on average, less than men, are more likely to care for children and the elderly, and are less likely to have health insurance and pension coverage in their jobs. In addition, since women generally live longer than men, they may be more concerned about building up retirement savings.

Lee hypothesized that dual-earner households (where women have more bargaining power) would be more likely to cut back spending in response to a stock market crash. Controlling for age, household size, and other factors, she found that dual-earner households were, in fact, more likely to reduce their spending in response to stock market losses and, among those who did cut spending, dual-earner households reduced spending by a larger amount than male-earner households. In fact, male-earner households often *increased* their spending after sustaining losses in the stock market.

Lee concluded that, "Higher female bargaining power causes house-hold-level decisions to be more aligned with female consumption preferences." This is a plausible theory that is supported by her data.

A random example

Suppose we had followed Lee's original strategy: look for correlations and see where the data take us, unencumbered by logic or theory. In order to demonstrate why good data scientists should stay off that path, we made up some completely random data.

First, we created 1,000 imaginary husband–wife households with bargaining power randomly distributed between 0 (complete male dominance) and 1 (complete female dominance). Then we looked at 100 household expenditure categories that are used in the calculation of the U.S. consumer price index (CPI); for example, cereal, women's footwear, and outdoor tools. For each of the 1,000 imaginary households, we used a random number generator to create fictitious data on percentage changes in spending in each of the 100 spending categories.

There are absolutely no systematic relationships between bargaining power and spending changes since these data were all created independently. Nonetheless, there might be coincidental correlations that would be uncovered by mindless data mining. And, indeed, there were. We calculated the correlation between bargaining power and spending changes for each of the 100 product categories. The average correlation was around zero since these data are pure noise, but there were several statistically significant correlations. Spending on small appliances and on canned fruits and vegetables were negatively related to bargaining power; spending on frozen fruits and vegetables and on land-line telephone services were positively related to bargaining power.

A data clown would either say "up is up," or make up a fanciful theory to explain these patterns. A data scientist would laugh and say, "That's why we need theories before looking at the data. Correlation is not causation."

Pop-Tarts and beer

For an example of a successful spending study, consider how Wal-Mart stocks its shelves when a hurricane is on the way. Customers don't just buy water and flashlights; they also buy strawberry Pop-Tarts and beer. Since historical data were analyzed, this appears at first glance to be data mining. It is actually analogous to a controlled experiment!

Recall that one major downside of data mining is the possibility of confounding variables. However, since hurricanes only affect a few stores out of many, Wal-Mart had a natural experiment that eliminates confounding influences like the day of the week or season of the year. This is almost as good as letting mad scientists randomly choose cities to be blasted by hurricanes and then comparing the shopping habits of the lucky and unlucky residents. The scientific method is alive and well at Wal-Mart.

Another problem with data mining is that correlation can get confused with causation. In this case, the relationship was between hurricanes and the products customers bought. It is highly unlikely that customers stocked up on Pop-Tarts in threatened cities for some reason other than hurricanes. Also, unless buying Pop-Tarts causes hurricanes, the relationship clearly goes in the other direction. The desire to prepare for a hurricane *caused* customers to stock up on beer and Pop-Tarts. We might not know exactly *why* people buy these products, just like ABTester didn't know exactly why some keywords are better than others. However, human common sense can be used to judge whether the correlation is sensible or coincidental. Pop-tarts don't have to be cooked, can be eaten any time of day, and practically last forever. The same is true of beer, though warm beer may be less satisfying than cold beer. In contrast, unfettered data mining might discover an unexpected, indeed inexplicable, correlation between hurricanes and, say, electric coffee makers.

Identifying useful relationships

There is a hierarchy of predictive value that can be extracted from data. At the top of the hierarchy are causal relationships that can be confirmed with a randomized and controlled experiment (ABTester) or a natural experiment (hurricanes and Pop-Tarts). Knowing a causal relationship is as close as you can get to true understanding.

Next best is to establish known or hypothesized relationships ahead of time and then test them and estimate their relative importance (IDx-DR). One notch lower are associations found in historical data that are tested on fresh data after considering whether or not they make sense (pregnancy and prenatal vitamins).

At the bottom of the hierarchy, with little or no value, are associations found in historical data that are not confirmed by expert opinion or tested with fresh data. Data scientists who use a "correlations are enough" approach should remember that the more data and the more searches, the more likely it is that a discovered statistical relationship is coincidental and useless.

The Seventh Pitfall of Data Science is:

Confusing Correlation with Causation

Being Surprised by Regression Toward the Mean

"Revenue is up 20%!"
"But I haven't done anything."
"Well, whatever you did worked."

After some initial confusion, the reality sank in. Revenue had gone up even though nothing had changed. If revenue can increase for no reason, how do we know that a revenue increase after we make changes was due to the changes? The company had been blindsided by the Eighth Pitfall of Data Science and the price had already been paid. Here's the story.

Jay worked for a company ("BestWeb") that specialized in maximizing ad revenue on web pages for themselves and for clients. Most domain names that can be purchased for under $10 are worthless, but someone who is good at identifying potentially profitable names can collect them and make some serious money.

BestWeb designed and conducted experiments that compared the current layout to as many as 20 alternative layouts across a million different domain names. When a user visited a client's website, a random-number generator sent her to one of the alternative layouts. BestWeb recorded a lot of data during the experiment, but the most important metric was RPM (revenue per mille) or revenue per 1,000 visitors.

The 9 Pitfalls of Data Science. Gary Smith and Jay Cordes. Oxford University Press (2019).
© Gary Smith and Jay Cordes 2019. DOI: 10.1093/oso/9780198844396.001.0001

These were well-designed experiments in that they tapped the power of random assignment. If BestWeb used a different layout each day, the results might be muddled by the nature of web traffic on the different days—for instance, people are typically more likely to click on ads on a Monday (yes, during work hours) than over the weekend. The only way to control for all the possible confounding influences is the approach they used: letting a random-number generator determine who is sent to which layout. After a few weeks of testing, BestWeb identified the most profitable layout.

Clients occasionally complained about "underperforming" domain names that they felt should be earning more. One of Jay's coworkers ("George") was given a list of domain names with revenue down over the past three months, and asked to tinker with the layouts to see if he could boost revenue. He was successful, spectacularly so. After he made changes, revenue always went up about 20 percent the next day. George gained a reputation as a rock star, but he and Jay warned that, technically, they should have a control group for comparison. They should tweak a random subset of the sites so that they could compare the revenue improvement to sites that had not been tweaked. Jay had a reputation for identifying data science pitfalls, but in cases like this, where clients were happy, managers were less receptive to his concerns. What are the chances that revenue would increase by 20 percent the day after George made changes, unless the changes really worked? What's the point of arguing with success?

Then, one day, George was too busy to make any changes. Revenue jumped up like it always did, and the account manager came by to congratulate George for yet another successful tweaking. Another account manager asked George to do the same thing for his clients. But now the game was up. George and Jay immediately realized what was going on because they had learned about *regression toward the mean*.

Gary had visited the company a few months earlier and explained a totally separate mystery that had vexed BestWeb. When they switched their default web-page layout to the winning layout in their experiments, the revenue increase after the change was generally smaller than the increase during the experiment. For example, if the winning layout showed a 5 percent increase in RPM during the experiment, the RPM would typically go up only 2 or 3 percent when they rolled it out after the experiment ended. What was wrong with their tests? BestWeb had spent weeks looking at different possible reasons for this frustrating pattern and couldn't find an explanation.

Gary said that this is exactly the sort of pattern they should anticipate when BestWeb chooses the best performing layout. It is called *regression toward the mean*, and it is endemic whenever data are used to identify the best performances: the best athlete, the best student, the best worker, the best medication, or the best web-page design.

To demonstrate the concept, Gary created a free-throw shooting contest for 20 experienced basketball players. Each player shot 10 free throws. The best performances were by Bill (who made all 10 shots) and Scott, Eric, and Cory (who each made 9 of 10). These seem to be the best four players and you might naively predict that if they took 10 more shots, Bill would again make 10 of 10, and Scott, Eric, and Cory would again make 9 of 10. However, when they did take 10 more shots, Bill only made 9. Cory made 10, but Scott only made 7 and Eric made 8. Three of the four best shooters in the first round did worse in the second round.

This is regression toward the mean. It occurs because there is chance, as well as ability, in free-throw shooting. The four players who made the most shots in the first round most likely benefited from good luck. It would be very unusual for someone to make 90 percent of his shots and say that he had bad luck. The top four players in the first round probably had good luck and, because luck is fleeting, they generally don't do as well in the second round.

Regression toward the mean is NOT the fallacious *law of averages*, also known as the *gambler's fallacy*. The fallacious law of averages says that things must balance out—that making a free throw makes a player more likely to miss the next shot; a coin flip that lands heads makes tails more likely on the next flip; and good luck now makes bad luck more likely in the future.

Gary once overheard the father of a Little League baseball player tell his son to lend his baseball bat to a poor hitter "to use up the bat's outs." This misguided father evidently believed in the fallacious law of averages—that outs must be balanced by hits. That's silly. Baseball bats have no memories or concerns about hits and outs. His son would have been better off practicing batting than loaning his bat to a bad hitter.

Jay once heard similar reasoning at a work meeting when a salesman told everyone that a sales guru had taught him that each *no* brings him closer to a *yes*, because 1 out of every 100 calls results in a sale. You just have to get through those 99 no's. Jay thought about it for a second and then piped up: "No, the calls are independent. You're not getting any closer to anything when someone says no." Everyone laughed at the cynical comment. No wonder Jay had the nickname Spock when growing up. If Gary had been at the meeting, he would have added, "If you get 99 *no*'s, maybe something's wrong with your sales pitch."

After losing election contests for the Governor of Washington in 2004 and 2008, Dino Rossi ran for the United States Senate in 2010. A *Washington Post* columnist wrote, "Is the argument for Dino Rossi basically that, by the law of averages, he has to win something eventually?" That's silly, too, though perhaps intentionally so. Losing elections doesn't make someone more likely to win the next election. If anything, it suggests that voters don't like this candidate. (Rossi lost the 2010 election, too, but, undeterred, ran for Congress in 2018—and lost again.)

Peter Bernstein, author of *Against the Gods*, gives this purported example of regression toward the mean: "Joseph had this preordained sequence of events in mind when he predicted that seven years of famine would follow seven years of plenty." That is the fallacious law of averages, not regression. If there is luck involved in famine and plenty, regression toward the mean predicts that an unusually large number of years of plenty will be followed by fewer years of plenty, not an equally large number of years of famine.

The principle behind regression toward the mean is that extraordinary performances exaggerate how far the underlying trait is from average. Bill, Scott, Eric, and Cory are all good free-throw shooters, just not as good as their first-round performance suggests. They are not destined to miss most of their shots in the second round in order to balance things out. They will probably continue to be above average in the second round, just not as far above average as they were in the first round.

Regression toward the mean also works for the worst performers. In the first round, two players made only 4 of 10 shots and two other players made 5 of 10. They are not very good shooters, but they are probably not as bad as they seem. They had a bit of bad luck and will probably do better in the second round. In fact, three of these four poor players did better in the second round. Figure 8.1 shows the results for all 10 players in these two rounds.

Regression toward the mean is a purely statistical phenomenon that has nothing at all to do with ability improving or deteriorating over time. The best free-throw shooters in the first round did not lose their touch, and

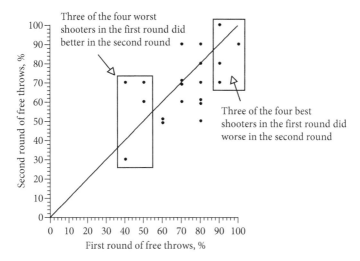

Three of the four worst shooters in the first round did better in the second round

Three of the four best shooters in the first round did worse in the second round

Figure 8.1 Performance regresses to the mean from the first round to the second round

the worst shooters did not suddenly learn how to shoot free throws. How do we know this? Because there is also regression toward the mean if we go back in time from the second round to the first round. Figure 8.1 showed that the best and worst shooters in the first round tended to be closer to average in the second round. Figure 8.2 shows that the best and worst shooters in the second round tended to be closer to average in the first round. Either way we look at it, forward or backward, the abilities of the best and worst performers are closer to the mean than are their performances, so their performances regress.

Gary explained that for each layout that BestWeb tests, the benefits fluctuate because of the luck of the draw (the random-number generator determines who goes to which page). To the extent there is randomness in their experiments—and there surely is—the actual benefits from the layout that was the most profitable in the experiment will probably be closer to the mean after the experiment, just as the best free-throw shooters in the first round didn't make as many shots in the second round. There is nothing wrong with that. BestWeb should still use the most successful layout. BestWeb just needs to recognize that it is perfectly natural for the benefits to be more modest than predicted by their tests.

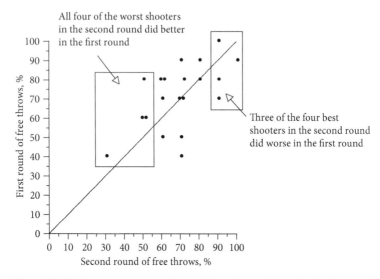

Figure 8.2 Performance regresses to the mean from the second round to the first round

Even after Gary had explained the concept to the employees of BestWeb, they still had to learn the hard way that they should anticipate regression toward the mean. When George was given a set of "underperforming" domains, everyone should have recognized that they were like those 40-percent free-throw shooters. Revenue wasn't likely to go down; it was bound to go up. Crediting George for the revenue increase is like asking the worst free-throw shooters in the first round change shirts and then attributing their improved performance to the new shirts.

Unleashing the Potential

The lesson sunk in for Jay and he decided to play a prank on his coworkers to ensure that they got it, too. Jay identified 3,000 domain names that were generating very little revenue. He told his coworkers that when he double-clicked on whatever image was in the upper left corner of each page, revenue increased by an average of 800 percent that day. He concluded that,

The results clearly indicate that double-clicking the image resets each domain and unleashes its true potential. I'll continue watching the results because it appears the effect is already wearing off, so perhaps a periodic refresh is in order. We can simply hire a temp to do it once every few weeks (man, my carpal tunnel is acting up). The domain list is below if you'd like to verify this miraculous result for yourself.

Some of Jay's coworkers looked at the unbelievable stats themselves, with one reporting back, "All right, I prepped the temp to handle this bi-weekly." Another was more concerned about the nonsensical technical aspects:

What is the "image in the top left corner" and why is it clickable? I don't think this a common feature in our designs. This sounds like a bug work-around to me, but I can't reproduce it as described.

Finally, one of the more skeptical analysts slowed down the burst of activity by saying, "I think Jay's messing with us." Jay fessed up. He hadn't double-clicked on any images. He hadn't done anything but collect the data and count on regression toward the mean to kick in. Out of a million domain names, he grabbed the ones with a tiny amount of revenue in the previous month. If a website is performing extremely poorly, it is much more likely to have had bad luck than good luck, which means that its future performance is likely to increase for no reason other than its performance had been improbably low.

This is a lesson that extends far beyond web-page pranks. Sometimes a company will experience a sharp downturn and bring in highly-paid management consultants to turn things around. The consultants poke around for a while, make some recommendations, and the company miraculously improves. That's Jay's joke on a larger scale. A company doing poorly is more likely to have been experiencing bad luck than good luck and, so, will probably do better in the future, whether or not the consultant's recommendations have any merit or, indeed, whether a consultant is even hired.

Seeing regression clearly

Howard Wainer, a statistician with the Educational Testing Service recounted the following story:

My phone rang just before Thanksgiving. On the other end was Katherine Tibbetts; she is involved in program evaluation and planning for the Kamehameha Schools in

Honolulu. Ms. Tibbetts explained that the school was being criticized by one of the trustees of the Bishop Estate (which funds and oversees the schools) because the school's first graders that finish at or beyond the 90th percentile nationally in reading slip to the 70th percentile by 4th grade. This was viewed as a failure in their education. Ms. Tibbetts asked if I knew of any longitudinal research in reading with this age group that might shed light on this problem and so aid them in solving it. I suggested that it might be informative to examine the heights of the tallest first graders when they reached the fourth grade. She politely responded that I wasn't being helpful.

Gary put this anecdote on a final examination and asked his students to, "Explain his suggestion to Ms. Tibbetts." Almost every student recognized this story as an example of regression to the mean. The tallest first graders were probably somewhat closer to average in fourth grade, just as the students with the highest scores in first grade tended to be closer to average in fourth grade.

However, one student came up with a creative, though misguided, answer: "Howard is suggesting that a change in the height of her students could make it difficult for the students to read assigned works if there is a greater distance between their eyes and the paper." Regression toward the mean is an elusive concept that can trick people into inventing far-fetched theories.

Ability and performance

We have loosely used the term "luck" to describe the chance variation that usually occurs when observed performances are used to measure unseen traits—when 10 free throws are used to assess shooting ability, when a few weeks of web traffic are used to measure a site's revenue potential, or when a company's recent performance is used to gauge its profitability. We are not saying that shooting ability, revenue potential, and profitability cannot be improved (or deteriorate), only that because performance changes can be due to the fleeting nature of luck, we should not automatically assume that the underlying traits are changing.

Consider the Scholastic Aptitude Test (SAT), where scores range from 400 to 1600. Each student's "ability" is the expected value of his or her score. The actual score might be above or below a person's ability, depending on such factors as the specific questions asked, guesses when the correct answer is unclear, and the student's health or mental state on the day of the test.

Regression toward the mean tells us that students who score well above the mean probably had more good luck than bad, and their ability is typically closer to the mean. Students who score well below the mean, on the other hand, probably had some bad luck and their ability is most likely closer to the mean.

How much closer to the mean? There is a remarkable formula, now known as Kelley's equation, which was derived in the 1940s by Truman Kelley, a professor at the Harvard School of Education. Kelley's equation says that the best prediction of a person's ability is a weighted average of the person's performance and the average performance of the group the person belongs to:

$$\text{estimated ability} = R(\text{performance}) + (1 - R)(\text{average group performance})$$

The term R is reliability, which measures the extent to which performances are consistent. If a group of students take two comparable tests, reliability is the correlation between their scores on these two tests.

If the test scores were completely random, like guessing the answers to questions written in a language the students don't understand, the reliability would be zero, and our best estimate of a person's ability would be the average score of the group.

At the other extreme, a perfectly reliable test would be one where some students do better than others, but each student's score is the same, test after test. Now, the best estimate of a student's ability is the student's score. This, too, makes sense.

For SAT tests, the reliability is about 0.9 and the average score is around 1100. Table 8.1 shows the estimated ability for various test scores. A student who has a score of 1100, which is the average score, is estimated to have an ability of 1100. As the scores move away from 1100, so do the estimates of ability, but by a smaller amount. A student with a score of 1400 is estimated to have an ability of 1370, while a student with a score of 800 is estimated to have an ability of 830.

A student's ability is the best predictor of the student's score if the SAT is taken a second time and there has been no change in ability. A student who scores 1300 is estimated to have an ability of 1280, so an unbiased prediction of this student's score on a second test is 1280. A regression to

Table 8.1 *Estimated ability for SAT scores with a mean of 1100 and a reliability of 0.9*

Score	Estimated ability
400	470
500	560
600	650
700	740
800	830
900	920
1000	1010
1100	1100
1200	1190
1300	1280
1400	1370
1500	1460
1600	1550

the mean mistake would be to think that scores measure ability perfectly, so that any change in scores must be due to changes in ability.

Paying for regression

High school students who are disappointed in their initial SAT scores often take expensive SAT-prep classes and are gratified when their scores improve. However, this may just be statistical regression toward the mean—especially if they felt that the initial scores were below their ability. Their scores could be expected to improve even if they had not taken a SAT-prep class.

An even more striking example was a report in *The New York Times* titled, "Drug May Help the Overanxious on S.A.T.'s." A study supported by the American Academy of Pediatrics identified 25 high school juniors who, based on IQ test scores and high school grades, had not done as well as predicted. An hour before they retook the SATs their senior year, they were given propranolol to relax them. Their scores improved, leading the researcher to conclude that, "Their parents and teachers had convinced them that, if they didn't do well on the SATs, they'd never get into college. The result was they approached the SATs with a tremendous amount of anxiety and fear." An obvious alternative explanation is regression toward the mean. These students were selected precisely because their initial SAT scores were lower than predicted, based on independent measures of their ability. We expect their scores to improve whether they take propranolol or just drink a glass of water.

A good data scientist would have used a control group and might have found that students taking propranolol before the SAT did worse than those taking placebos, since one side effect of propranolol is drowsiness. Unfortunately, many well-intentioned parents reading the *Times* story may have fallen into that pitfall.

When will regression toward the mean not occur? There will be no regression if there is no chance involved. For example, if you measure Bill's age (and Bill tells the truth), there is no variation in the data and no direction in which to regress. With data that remain consistent and predictable, there's no noise obscuring the signal. However, most data are not like that.

Consider Sir Francis Galton's bean machine in Figure 8.3. As beans are dropped into the top of the machine, they bounce around until settling into a nice bell curve at the bottom. Think of the data you see in the world as the beans collecting at the bottom, and your job as a data scientist is to predict where the next bean will fall. You're not allowed to peek at the top of the bean machine to see where they are coming from; you can only examine the distribution of the beans that have already fallen. Suppose you see a bean that lands some distance from most of the previous beans. Intuitively, you know better than to predict that the next bean will fall on top of the last bean. You will not be surprised by the regression toward the beans.

Don't buy a portfolio from a devious data analyst

At BestWeb, Jay collected other data to use as a warning for people who buy and sell existing domains. Generally, potential buyers request data that include the last three months of revenue. Figure 8.4 is a chart of the average daily revenue for the domains that Jay had selected. He sent out an email to managers, predicting that the revenue would plummet next month.

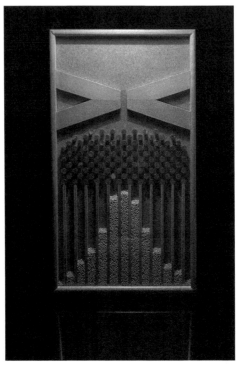

Figure 8.3 Sir Francis Galton's bean machine. Credit: Matemateca (IME/USP)/Rodrigo Tetsuo Argenton

They must have thought he was crazy; revenue is clearly trending upwards. If these domains were for sale, the purchase price would be negotiated as a multiple of the most recent month's revenue, which should turn out to be a bargain price, given how much higher revenue will be in a few months!

Jay sent out a follow-up email a few months later that included Figure 8.5, which confirmed his prediction.

There's clearly something strange going on here. How could Jay possibly know that revenue would drop? Figure 8.6 shows revenue for the entire year, and reveals the trickery.

First of all, we should emphasize that, as crazy as Figure 8.6 looks, this is real revenue for an actual collection of domain names. Jay selected the

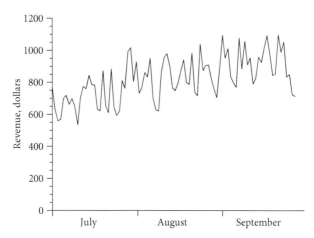

Figure 8.4 Revenue for the last 90 days, looking good!

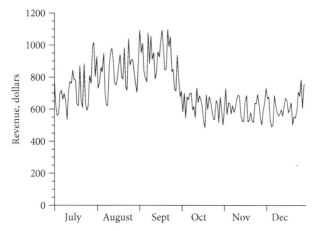

Figure 8.5 What happened?

domain names strategically in order to show indisputable regression toward the mean. First, he identified domain names that had very low revenue in May–June; then he picked a subset of *that group* that had the highest revenue over the following three months (July–September), and were likely to regress. Because of the inevitable volatility in revenue and

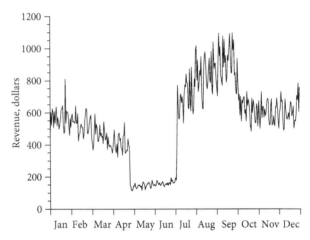

Figure 8.6 Revenue for the entire year

the large number of domains to choose from, the regression is truly dramatic. As with free-throw shooters, we can see the regression going both forward and backward in time relative to the July–September pop.

Jay exploited the fact that revenue fluctuates around an average—nothing more or less. He simply chose domain names that were performing poorly in May through June and doing well in July through September. He was p-hacking. When the October–December data came in, we see a more realistic revenue estimate of around $600 per day.

Clients often thought portfolio sellers were ripping them off because the collection of names they bought experienced a sudden decline in revenue after the sale was finalized. The Eighth Pitfall of Data Science warns buyers to expect that when they buy the best-performing domain names. If, instead, they bought the worst-performing domain names, they would likely be pleasantly surprised when revenue regresses upward after the purchase.

Baseball players

Sometimes, when trying to explain regression toward the mean to his coworkers, Jay would use baseball statistics. One time, the rebuke came back: "Domain names aren't baseball players," to which Jay responded "in what way?" Practically any set of data is very much like baseball statistics. There is ability, there is uncertainty, and performances fluctuate around ability. This isn't a conjecture; anyone who looks at baseball statistics will see that it is true. In fact, let's create a "regression team," but with baseball players instead of domain names.

The major league baseball players shown in Table 8.2 and Figure 8.7 are the players with the top 10 batting averages in August 2017 who had at least 50 at-bats. By selecting players with a great August, you can bet that they will have a worse September, and they did—they regressed from a 0.378 batting average in August to 0.283 in September. They also regressed the previous month, from August to July, because they were luckier in August than they were in July and September.

If you want to impress your friends by pretending that you have a crystal ball, write down the names of the ten players with the top batting averages for the month and predict that their average batting average will be substantially lower next month. If you show them your players' batting averages for the month before the current month, they will be even more skeptical of your claim—since the players seem to be on an upward trend. Feel free to repeat this process every month until your friends are convinced you can see the future. You can also select the

Table 8.2 *Regression in baseball batting averages, 2017*

Player	Team	July	August	September
Garcia, A.	Chicago White Sox	0.216	0.423	0.355
Beckham, T.	Baltimore	0.160	0.394	0.410
Vazquez, C.	Boston	0.254	0.385	0.277
Blackmon, C.	Colorado	0.370	0.383	0.286
Polanco, J.	Minnesota	0.078	0.373	0.267
Castillo, W.	Baltimore	0.091	0.371	0.170
Andrus, E.	Texas	0.232	0.368	0.266
Villar, J.	Milwaukee	0.217	0.364	0.280
Inciarte, E.	Atlanta	0.269	0.362	0.272
Herrera, O.	Philadelphia	0.360	0.360	0.244
Average		0.225	0.378	0.283

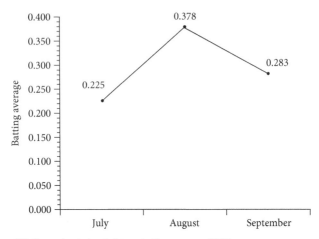

Figure 8.7 Regression in baseball team batting averages, 2017

players with the worst batting averages for the month and predict that they'll improve. Baseball players and domain names are the same thing, after all.

It can't fail

The reason that we've dedicated an entire Pitfall of Data Science to regression toward the mean is not just because it is counterintuitive. It is because it shows up time and time again and can lead to costly errors.

A manager at BestWeb told Jay that they wanted to test a new concept on domain names that weren't making any money, figuring there's nothing to lose.

"Great, just make sure to test it on a random half of the names." Jay responded.

"Why? They're not making any money!"

"Exactly. They can't do any worse, so the test can't fail."

Jay knew that domain names that didn't make money one week often make money the next week. Without a comparison set, the "experiment" can only result in success and, therefore, isn't a true experiment. Revenue will probably go up, but maybe less than it would have if the domain names had been left alone. When it comes to science, saying "it can't fail" is a bad thing, not a good thing. Experiments need the possibility of failure or there's nothing to learn from them.

Carrots versus sticks

A wonderful example of regression toward the mean is when Daniel Kahneman, a Nobel Laureate in Economics, tried to convince Israeli flight instructors that trainees should be praised instead of punished. A senior instructor argued that Kahneman had it backwards:

On many occasions I have praised flight cadets for clean execution of some aerobatic maneuver, and in general when they try it again, they do worse. On the other hand, I have often screamed at cadets for bad execution, and in general they do better the next time. So please don't tell us that reinforcement works and punishment does not, because the opposite is the case.

Do pilots who are praised become complacent, while those who are screamed at try harder?

Kahneman tried to demonstrate the senior instructor's error by drawing a target on the floor and asking each instructor to turn around and throw two coins, one after the other, at the target. The officers who were closest on their first throw generally did worse on their second throw, while those who were farthest away on their first throw usually did better on their second throw. Did the first-throw winners become complacent, while the first-throw losers tried harder? No, because they didn't look at their first throws before making their second tosses. In addition, if we look at the tosses in reverse order, those with the best second throws generally did not do as well on their first throws, while those with the worst second tosses typically did better on their first throws.

The answer to this seeming paradox is that these coin throws were just like two rounds of basketball free throws. Those officers who did well on either the first or second throw generally had good luck in the angle and force with which they threw the coin and the way the coin bounced and rolled after landing, but, as a Swedish proverb states: "luck does not give, it only lends."

In the same way, the pilots who had the best initial flights are generally not as far above average in ability as they were in performance, and will usually not do as well on the next flight. Similarly, the pilots who had the weakest flights are probably not as incompetent as they seem and are likely to do better next time even if the instructor screams into a pillow instead of at the pilots.

Regression toward the mean is everywhere

It is not just Internet experiments, flight tests, basketball shooting, and coin throwing. Regression toward the mean is all around us, and likely to be overlooked by inattentive data scientists.

In educational testing, many schools use test scores to identify students who need special tutoring, and then measure the success of the program by seeing whether these low-scoring students improve after the tutoring sessions. Regression toward the mean teaches us that the students who did the worst on the initial tests can be expected to do better on subsequent tests even if the tutors do nothing more than wave a hand over their heads.

Massachusetts public schools were given improvement goals for the year 2000 based on their 1999 average scores on standardized tests. There was considerable consternation when most of the schools with the lowest scores in 1999 met their goals while many of the schools with the highest scores did not; in fact, their scores often fell. It seemed that weak students were learning and strong students were unlearning. The Eighth Pitfall of Data Science teaches us to expect this.

In medical testing, most medical treatments that worked well enough to get government approval and/or be published in prestigious medical journals turn out to be less effective in practice than they were in the experimental tests. This pattern is so common, it even has a name—the "decline effect." Some researchers who have seen the decline effect firsthand with their own research are so perplexed that they go off on wild-goose chases (just like BestWeb did) looking for a causal explanation, when the reason is right in front of them. If the initial positive results were partly due to chance—for example, random differences in the health of the people selected for the control group and treatment group—it should be no surprise that the results regress toward the mean. This is why treatments should not be recommended based on just one study. Science makes progress when independent studies fit together and reinforce each other.

In nutrition studies, suppose that physical examinations are given to a large number of people and those with the highest cholesterol readings are identified and given a special diet. We can expect their cholesterol readings to improve even if the dietary instructions are nothing more than, "Talk to your food before eating it."

One medical study concluded that the placebo effect is often regression toward the mean. Patients who are feeling poorly are sometimes given worthless treatments (even sham surgeries!) and report feeling better afterward. This "placebo effect" might be nothing more than regression toward the mean. You can skip worthless treatments and sham surgeries, and you'll probably feel better in a few days.

There is regression in inherited traits such as height, weight, and intelligence because the things we measure are noisy indicators of the genetic influences that pass from one generation to the next. Regression toward the mean tells us that unusually tall parents generally have shorter children, and exceptionally tall children typically have shorter parents. Unusually athletic parents usually have less athletic children, while exceptionally athletic children are generally more athletic than their parents. (Don't expect the son of Steffi Graf and Andre Agassi to be the best tennis player in history.) A regression fallacy would be to assert that parents who are intellectually or athletically gifted stunt their children, or that parents who are intellectually or athletically challenged nourish their children.

We see a golfer win the British Open and conclude that he is the best golfer in the world. We see a student get the highest score on a test and conclude that she is the best student in the class. We see a medical test result and conclude that a patient has a disease.

If the British Open champion loses the next tournament, we might conclude that he wasn't focused. If the student with the highest test score does not do as well on her next test, we might conclude that she did not study. If the patient with the worrisome medical result fares better a month later, we might conclude that whatever treatment was prescribed was effective.

If, instead, we recognize that luck may have played a role, we are more likely to realize that the tournament winner is not necessarily the best player, that the student with the highest test score is not necessarily the best student, and that the patient with a worrisome medical reading on one test does not necessarily have a disease.

We will understand that several golfers are good enough to win a tournament and several students are good enough to get the highest test score, and they take turns doing so—not because their abilities fluctuate week to week, but because their luck comes and goes. We will understand that medical test results can fluctuate even if the patient's condition does not.

Being aware of regression toward the mean can even bring us peace of mind. Jay experienced this when a routine medical test came back positive for something alarming. Knowing that the more surprising a result, the less likely it is to be true, he decided not to worry about it unless it was confirmed by a follow-up test. A more reliable test was ordered to confirm the diagnosis and sure enough, it came back negative.

In sports, a player or team that accomplishes something exceptional most likely benefited from good luck and will subsequently regress. It is a mistake to believe that champions choke or players are jinxed by winning a rookie-of-the-year award or by appearing on the cover of *Sports Illustrated*.

After Oklahoma won 47 straight college football games, *Sports Illustrated* ran a 1957 cover story proclaiming, "Why Oklahoma is Unbeatable." Oklahoma lost its next game, 7 to 0 to Notre Dame, and people started noticing that other athletes who appear on the cover of *Sports Illustrated* are evidently jinxed in that they do not perform as well afterward. In 2002, *Sports Illustrated* ran a cover story on the jinx with a picture of a black cat and the wonderful caption "The Cover No One Would Pose For." More recently, we have the Madden Curse, which says that the football player whose picture appears on the cover of *Madden NFL*, a football video game, will not perform as well the next season.

The *Sports Illustrated* jinx and the Madden Curse are extreme examples of regression toward the mean. When a player or team does something exceptional enough to earn a place on the cover of *Sports Illustrated* or *Madden NFL*, there is essentially nowhere to go but down. To the extent luck plays a role in athletic success, and it surely does, the player or team that stands above all the rest almost certainly benefitted from good luck— good health, fortunate bounces, and questionable officiating. Good luck cannot be counted on to continue indefinitely, and neither can exceptional success.

It is also a mistake if you believe that athletes are jinxed by a teammate or announcer talking about how well they are doing, or by a fan watching them on TV:

[F]ans of the Cleveland Browns had to cringe on Sunday when CBS put up a graphic noting their team hadn't turned the ball over in 99 trips to the red zone. They had to mutter when the broadcasting team of Andrew Catalon and Steve Beuerlein praised the Browns for their error-free ways. And they definitely had to be cursing as running back Isaiah Crowell immediately coughed up the ball just seconds later.

Couldn't you have left well enough alone, asked Cleveland?

The "broadcaster jinx" is regression toward the mean plus selective recall. No team has a zero-percent probability of fumbling, so a zero-percent fumble rate exaggerates their ability not to fumble. In addition, we remember when announcers say something that is immediately contradicted and forget the times when they are not contradicted.

Since regression toward the mean goes both ways, it can also appear as an anti-jinx. This happened recently when Jay's brother messaged him wondering when Klay Thompson, a National Basketball Association (NBA) player, would get out of his shooting slump. A few hours later, Thompson broke the all-time NBA record for three-pointers in a game. Maybe Thompson should thank Jay's brother for the magic spell?

Why are movie sequels usually not as good as the originals? It's because sequels are made of movies that are unusually good. Many things contribute to a movie's success, and luck is one of those things. If you want movie sequels to be better than the originals, there's an easy fix: make sequels to the bad ones.

There is regression in the stock market because investors overreact to corporate news. Unusually strong or weak earnings tend to regress and the most optimistic and pessimistic earnings forecasts are generally too extreme. So stock prices are often too high for companies whose earnings have increased dramatically or are predicted to do so, while the reverse is true of companies with weak earnings or pessimistic forecasts. When earnings turn out to be closer to average than they have been or were predicted to be, stock prices adjust. In the same way, stocks that drop sharply are often better investments than are stocks that surge, and stocks that are removed from the Dow Jones Industrial Average are generally better investments than the stocks that replace them.

There is regression in the evaluation of job candidates, no matter whether they are clerks, CEOs, or politicians. Whenever there is uncertainty about how well a person will do on the job, those who seem the most qualified will most likely not do as well as anticipated. As Warren Buffett put it, "When a manager with a great reputation meets a company with a bad reputation, it is the company whose reputation stays intact."

OCT 29, 9:21 AM

Little Steph! I hope Klay
finds his shot again soon.

OCT 29, 8:41 PM

This happened after you
sent your message right,
ha ha?

Klay Thompson drilled 🔢🔢 triples to
BREAK the record for most 3-point field
goals in a game!

Catch all of his best plays from his historic
night as he tallies 52 PTS in just 27
minutes of action.
NBA

Figure 8.8 Behold the anti-jinx!

Running for president is the ultimate job application, and, yes, there is regression. Since 1937, the Gallup polling organization has been asking Americans, "Do you approve or disapprove of the way —— is handling his job as president?" Every person elected president since 1937 has had lower favorability ratings at the end of their first term than at the beginning. The presidential candidates who win the election are seldom as good as they seemed at the time.

It is also true of the search for soul mates. A frustrated woman once sought help from the advice column "Ask Amy":

I keep meeting men who appear to have it all together. Before I know it—it is revealed that they do not. Oftentimes they live with relatives, they are hung up on an ex-wife or girlfriend, they are often financially irresponsible, and/or have serious emotional issues.

She's hardly alone. Believe it or not, our frustration and disappointment may just be another example of regression toward the mean.

Teachers don't know who will be their best students. Doctors don't know which treatments will be the most effective. Businesses don't know which new products will be the most successful. National Football League teams don't know which college draft picks will perform the best in the pros. Companies don't know which job candidates will be the best employees. Voters don't know which politicians will be the most effective once elected. And, yes, we don't know which prospective soul mates will be genuine and which will be duds.

When we try to assess a person, place, or thing, we may overrate it or underrate it. The regression insight is that those things we rate the highest are more likely to be overrated than underrated. When a student scores 95 on an initial test, how likely is it that this student had a bad test and would normally score even higher? Not likely at all, because there are many more students who would normally score below 95 than above 95. When a job candidate aces an interview, how likely is it that this person usually does even better? When a company's earnings increase by 30 percent, how likely is it that the company had a bad year?

Thus, the student, job candidate, and company all regress. The student will probably get a somewhat lower score on the next test; the job candidate will probably not perform as well on the job as expected; and the company's earnings will probably increase by less than 30 percent next year.

Regression toward the mean is the key to the universe.

Living with regression

The logic of regression is simple, but powerful. Our lives are filled with uncertainties. The difference between what we expect to happen and what actually does happen is, by definition, unexpected. We can call these unexpected surprises chance, luck, or some other convenient shorthand. The important point is that, no matter how reasonable or rational our expectations, things sometimes turn out to be higher or lower, larger or smaller, stronger or weaker than expected.

We are predisposed to discount the role of luck in our lives—to believe that successes are earned and failures deserved. We misinterpret the temporary as permanent and invent theories to explain noise. We overreact when the unexpected happens, and are too quick to make the unexpected the new expected. The key to understanding regression toward the mean is to look behind the data—to recognize that when we see something remarkable, luck was most likely involved and, so, the underlying phenomenon is not as remarkable as it seems.

Whenever there is uncertainty, people often make flawed decisions due to an insufficient appreciation of regression toward the mean. Don't be one of them.

The Eighth Pitfall of Data Science is:

Being Surprised by Regression Toward the Mean

Doing Harm

> For artificial intelligence to be truly smart it must respect human values—including privacy. If we get this wrong, the dangers are profound.
>
> —Tim Cook

In 2013, Eric Loomis was arrested in La Crosse, Wisconsin, and charged with five counts related to a drive-by shooting: attempting to flee a traffic officer, operating a motor vehicle without the owner's consent, recklessly endangering safety, possession of a firearm by a felon, and possession of a short-barreled shotgun or rifle. He denied being involved in the drive-by shooting, but pleaded guilty to the less severe charges of attempting to flee a traffic officer and operating a motor vehicle without the owner's consent.

The Presentence Investigation Report (PSIR) included the conclusions from the COMPAS (Correctional Offender Management Profiling for Alternative Sanctions) computer model created by a company named Equivalent (formerly Northpointe). The COMPAS model estimates recidivism risk based on 137 inputs gleaned from interviews and public information about defendants.

We know very little about the specific inputs or how much importance they are assigned by COMPAS because it is a proprietary black-box algorithm. Equivalent sells its assessments, and argues that it cannot divulge the details behind the assessments, because the value of their algorithm would be undermined by copycat competitors.

The 9 Pitfalls of Data Science. Gary Smith and Jay Cordes. Oxford University Press (2019).
© Gary Smith and Jay Cordes 2019. DOI: 10.1093/oso/9780198844396.001.0001

We do know some interview questions and it is not comforting:

How many of your friends/acquaintances are taking drugs illegally?
If people make me angry or lose my temper, I can be dangerous.

Seriously, who would answer, "Dozens" and "Certainly"? Risk-assessment algorithms surely ignore the answers to questions that any sensible criminal would lie about. They must be based on more subtle data, but what data?

Equivalent acknowledges that its algorithm is intended for groups of people who are statistically similar and says that it does not make specific predictions for individuals, like Mr. Loomis. The PSIR for Loomis specifically stated that, "risk scores are not intended to determine the severity of the sentence or whether an offender is incarcerated." But judges who are awed by computers are likely to be persuaded by computer recommendations, despite disclaimers that read like legal boilerplate that is not to be taken seriously.

The circuit court judge explicitly referenced COMPAS in his decision to sentence Loomis to six years in prison:

You're identified, through the COMPAS assessment, as an individual who is at high risk to the community.

In terms of weighing the various factors, I'm ruling out probation because of the seriousness of the crime and because your history, your history on supervision, and the risk assessment tools that have been utilized, suggest that you're extremely high risk to re-offend.

Loomis appealed his six-year sentence, based in part on the argument that his constitutional right to due process was violated because there was no practical way for him to challenge the validity of a black-box algorithm. Were some inputs inaccurate? Were some weights inappropriate? Did they enter some data for a different Eric Loomis? Did the algorithm put an unreasonably large weight on some extraneous factor, like the color of the car he drives? There is no way to know.

The Wisconsin Supreme Court rejected Loomis' appeal and lauded evidence-based sentencing—using data instead of subjective human opinion. Better to rely on an impartial computer program than on possibly biased judges.

The Court argued that although Loomis is not able challenge how the "algorithm calculates risk, he can at least review and challenge the resulting risk scores." How can he do that, other than by creating his own algorithm? And even if Loomis had his own algorithm, how would a court choose between competing black boxes?

Loomis' appealed to the Supreme Court of the United States. The Wisconsin Attorney General argued that the Court should adopt a wait-and-see attitude: "The use of risk assessments by sentencing courts is a novel issue, which needs time for further percolation." In the meantime, Loomis "was free to question the assessment and explain its possible flaws." On June 26, 2017, the Supreme Court declined to hear the case.

Wisconsin is hardly alone. Many states now use risk-assessment algorithms and the Model Penal Code (Final Draft; American Law Institute, 2017) endorses the use of data-driven risk-assessment models.

How did we come to come to this point where people are imprisoned based on inscrutable black-box models, where people languish in prison while algorithms percolate?

Too many people believe that computers are smarter than humans and, so, we should do what computers tell us to do—even if we have no idea why a program recommends this particular action in this specific case.

Algorithmic criminology is increasingly common in pre-trial bail determination, post-trial sentencing, and post-conviction parole decisions. One developer wrote that, "The approach is 'black box,' for which no apologies are made." He gives an alarming example: "If I could use sunspots or shoe size or the size of the wristband on their wrist, I would. If I give the algorithm enough predictors to get it started, it finds things that you wouldn't anticipate." Things we don't anticipate are things that don't make sense, but happen to be coincidentally correlated.

Some predictors may well be proxies for gender, race, sexual orientation, and other factors that should not be considered. People should not have onerous bail, be given unreasonable sentences, and be denied parole because of their gender, race, or sexual orientation—because they belong to certain groups. What should matter are the specific facts of a particular case.

We would need data for hundreds, perhaps thousands, of cases in order to provide evidence that a particular algorithm is discriminatory and, even then, the developers would, no doubt, say that they are constantly "revising and improving" their algorithm, so past results are irrelevant.

It has been estimated that COMPAS has a 65 percent success rate for predicting whether a defendant who is released will commit another crime within two years. Two Dartmouth researchers looked at the same database and found a 67 percent accuracy using predictions based on just two factors: age and the number of prior convictions. Duh. Young people who have committed a lot of crimes are likely to commit more crimes.

A study by Duke University researchers obtained similar results using sex, age, and prior convictions. Maybe that is about as good as it gets when predicting recidivism. Motives, emotions, and circumstances that are hard to anticipate, let alone quantify, make it very difficult to predict human behavior with a high degree of accuracy. If age and prior convictions are about as good as it gets, let's not rely instead on black-box algorithms that make predictions based on rubbish like wristband sizes.

The allure of AI

Artificial intelligence works wonderfully at many specific, narrowly defined tasks such as recommending music, movies, or products based on a statistical analysis of what we have listened to, watched, or purchased in the past and the behavior of other people who have done similar or dissimilar things.

If a 75-year-old buys a bird feeder, he might be in the market for bird food, another bird feeder, or a wind chime if that's what other bird-feeder buyers tend to purchase. He might not be interested in jet skis. That makes sense.

If we ask Siri, Alexa, or Google Assistant an obscure question and it gives a seemingly correct answer, we are (and should be) impressed. If it tells a joke, we might laugh:

Question: Which cell phone is best?
Siri: Wait…there are other phones?
Question: What is zero divided by zero?
Answer: Imagine that you have zero cookies and you split them evenly among zero friends, how many cookies does each person get? See, it doesn't make sense and cookie monster is sad that there are no cookies, and you are sad that you have no friends.
Question: Do you have a boyfriend?
Siri: Why? So we can get ice cream together, and listen to music, and travel across galaxies, only to have it end in slammed doors, heartbreak and loneliness? Sure, where do I sign up?

Those who think computers are clever, based on such responses, might not realize that these answers were scripted by humans. Google Assistant does the same, even using writers from Pixar and *The Onion* to help craft clever responses.

Computers can tell us the square root of any number, the capital of any country, and the directions to the nearest gas station. In addition to impressive everyday tasks, computers have had several highly publicized successes, like winning at Jeopardy!, chess, and Go, further enhancing the belief that computers are smarter than humans. The reality is that these programs perform spectacularly well on narrowly defined tasks that have clear goals, but are not really *thinking* in any normal sense of the word.

Jeopardy!

The Watson program, created by IBM to play Jeopardy!, was an astonishingly powerful search engine capable of finding words and phrases quickly in its massive database, and it had an electronic trigger finger that pushed its buzzer faster than its human competitors. It was perfect with factoids ("Ronald Reagan's middle name"—"What was Wilson?") but struggled with puns and double meanings ("Sink it and you've scratched"—"What is the cue ball?"). The head of IBM's Watson team stated the obvious: "Did we sit down when we built Watson and try to model human cognition? Absolutely not. We just tried to create a machine that could win at Jeopardy."

In 2011, Watson defeated two of the best all-time Jeopardy! champions. In the end, it had a higher score than both humans put together. The machine itself was a beast: 2,880 parallel processors, each cranking out 33 billion operations per second, and had 16 terabytes of RAM. There were some humorous mistakes, such as the time Watson considered milk to be a nondairy powdered creamer. However, Watson's victory was stunning.

Backgammon

When world champion Luigi Villa lost a backgammon match 7–1 to a computer program written by Hans Berliner in 1979, it was the first time that a machine had beaten a world champion in any game. Later analysis showed that the human was actually the stronger player and only lost due to bad luck, but Pandora's box had been opened. TD-Gammon was developed in 1991 and was followed in 1992 by Jellyfish and Snowie. There were no databases of moves and no expert advice given to the machines. They were told only which features of the board may be important (like the number of consecutive blocking points). They played millions of games against themselves, following a simple strategy at first, but then discovering how to value board positions properly and find moves that helped improve their positions. The algorithms had taught themselves to play backgammon expertly!

It wasn't long before the machines were as good as the best players in the world. The final insult to humanity was when a backgammon player named Jeremy Bagai wrote a book that used Snowie to identify mistakes in the classic strategy guides written by human experts. Bagai used a computer program to re-examine the bibles of backgammon and correct them.

Checkers

In 1989, a team led by Jonathan Schaeffer from the University of Alberta created a computer program called Chinook that could play checkers. In 1990, Chinook was ready to take its first crack at the world title, but fell short against Marion Tinsley. Tinsley, who is considered the best checkers player of all-time, won 4 games to Chinook's 2, with 33 draws. In the rematch in 1994, it seemed that Chinook might actually have a chance against the seemingly unbeatable human champion. However, after 6 draws, the match came to an unfortunate and premature end: Tinsley had to concede due to abdominal pains, later diagnosed as cancerous lumps on his pancreas.

Using impressive-sounding strategies such as minimax heuristic, depth-first search, and alpha-beta pruning, in combination with an opening-move database and a set of solved endgames, Chinook held on to its title with a 20-game draw against the #2 player, Don Lafferty, but hadn't yet truly become unbeatable. During the match, Lafferty broke Chinook's 149-game unbeaten streak, which earned him the title, "last human to beat a top computer program at checkers."

After the next match, in 1995, it was official: machine had surpassed man. Don Lafferty fell by a score of 1–0 with 35 draws. A couple years later, Chinook retired after being unbeaten for three years. In 2007,

Schaeffer's team announced that checkers had been solved. They proved that checkers is a draw if both sides play perfectly. Machines would never lose at checkers again.

Chess and Go

A company called DeepMind created AlphaGo, which did something that had been considered impossible—defeat the world's best Go players. AlphaGo trained on tens of millions of historical Go matches, as well as games played against itself, and then beat the best humanity had to offer. Using neural networks, AlphaGo was able to identify winning strategies without searching the impossibly large tree of possible moves. It sometimes made mysterious moves never before made by human experts. AlphaGo's highest profile win was its 4–1 victory over top-ranked professional player Lee Sedol in 2016.

As if that weren't impressive enough, a year later, DeepMind introduced AlphaGo Zero. Unlike AlphaGo, this version learned the game by playing itself (zero human games were necessary). After 3 days of training, it beat AlphaGo 100 games to 0.

Then came AlphaZero, another deep neural network algorithm that was trained by playing games against itself. This more general algorithm could play chess and shogi, as well as Go. The results were stunning. In Go, after 34 hours of training, it beat AlphaGo Zero 60–40; in shogi, after 12 hours of training, it beat a world champion program 90–8 (with 2 draws); and in chess, with 9 hours of training, AlphaZero collected 25 wins as white, 3 as black, and 72 draws against a world champion chess program.

As with the backgammon program Snowie, AlphaZero sometimes makes moves that puzzle human grandmasters and cannot be explained by the developers. Demis Hassabis, the CEO of DeepMind, says that, in one game, AlphaZero moved its queen to a corner of the board, contradicting the human wisdom that the queen, the most powerful chess piece, becomes more powerful in the center of the board. In another game, AlphaZero sacrificed its queen and a bishop, which humans would almost never do unless there was an immediate payoff. Hassabis said that AlphaZero "doesn't play like a human, and it doesn't play like a program. It plays in a third, almost alien, way."

In addition to trouncing human chess players, computer algorithms have created an "endgame tablebase" that contains worked-out solutions for all chess positions with fewer than 8 pieces on the board. It may not sound like much, but it requires 140,000 GB of storage. We'll have to wait a while for the 8-piece tablebase, which is estimated to require 100 times as much storage. The tablebases are created by working backwards from checkmate positions and recording whether each possible prior position is a win, loss, or draw, assuming perfect play. It also stores the minimum number of moves it would take to reach that result.

The endgame tablebase contained some big surprises for humanity. Some positions which chess masters believed to be draws (such as queen and bishop versus two rooks) turned out to be winnable. Some of the checkmates require over 500 perfect moves (see Figure 9.1), which not only dwarfs the deepest analysis ever done by humans, but also challenges the 50-move rule which allows a player to claim a draw if there have been 50 consecutive moves without a capture or pawn move. Big data is disrupting chess to the point where people are debating whether the rules should change!

Tim Krabbé described his experience studying some of these perfect endgames:

A grandmaster wouldn't be better at these endgames than someone who had learned chess yesterday. It's a sort of chess that has nothing to do with chess, a chess that we could never have imagined without computers. The [tablebase] moves are awesome, almost scary, because you know they are the truth, God's Algorithm – it's like being revealed the Meaning of Life, but you don't understand one word.

Figure 9.1 Can you find the mate in 546 moves? Photo by Jacqueline

Should we welcome our new computer overlords?

We don't know *how* AlphaZero and AlphaGo select their moves, but we do know that they are terrific. Is winning at chess and Go the same as doing science? There is an important distinction between Abstract-World and Real-World.

Some games, like tic-tac-toe and checkers, can be "solved" in that there is an optimal strategy that always works. The moves can be proven perfect by iterating through all possible outcomes, just as mathematical theorems can be proven. This is Abstract-World, where machines are the undisputed champions.

Real-World data, in contrast, are messy, uncertain, and possibly biased. From a data scientist's point of view, the challenge is to make models that are robust in that they are not led astray by unimportant real-world details. Suppose that instead of being trained on a precise numerical representation of board positions, AlphaGo had been trained on images of boards and pieces with different colors, sizes, and shapes. Such real-world features would make it much more difficult for a computer to determine a winning combination of moves, though humans would tune out such distractions.

Machines do not know which features to ignore and which to focus on, since that requires real knowledge of the real world. In the absence of such knowledge, computers focus on idiosyncrasies in the data that maximize their success with the training data, without considering whether these idiosyncrasies are useful for making predictions with fresh data. Because they don't truly understand Real-World, computers cannot distinguish between the meaningful and the meaningless. Unlike Abstract-World, Real-World is constantly changing, so AI systems trained on the past may be worthless in the present. Images of tanks may no longer contain clouds.

This is not to say that these algorithms will never be useful for scientific discovery. DeepMind has already stunned the scientific community with AlphaFold, an AI approach that crushed almost 100 competitors in predicting the 3D shapes of proteins. It correctly predicted the structure of 25 out of 43 proteins, while the second-place competitor predicted only three.

Despite their freakish, superhuman skill at board games, computer programs do not possess anything resembling human wisdom and common sense. These programs do not have the general intelligence needed to deal with ill-defined situations, vague rules, and ambiguous, even contradictory, goals. Humans imagine, while computers remember.

Deciding where to go for dinner, whether to accept a job offer, or who to marry is very different from recognizing that a certain chess position is a checkmate, which is why it is perilous to trust computer programs that don't understand the world to make decisions for us, no matter how well they do at board games.

Job applications

Computer algorithms are increasingly being used to evaluate job applicants, even though the words and numbers put in a job application may have little to do with how well a person can do the job. Laszlo Bock, Google's Vice President of People Operations, argued that the most important questions are

"Give me an example of work you did that is exactly like the work you're going to do." They're called "structured behavioral interview questions." Everything else is kind of a waste of time.

Jane Swift, a senior manager at another firm, told Gary that if she had two job applicants, one with a computer science degree from a top university and the other with no college degree who had spent the last two years doing—and doing well—what she was looking for, she would hire the second candidate.

It's like the lyrics in the song "Them That Got":

That old sayin' them that's got are them that gets
Is somethin' I can't see
If ya gotta have somethin'
Before you can get somethin'
How do ya get your first is still a mystery to me

Gary asked Jane how a person with no experience could get the experience needed to be hired by her. Jane's answer was immediate. Find an internship doing what you want to do for a company you want to work for. Offer to work for free if that's what it takes to get the internship. Once the company sees that you are reliable, meet deadlines, do excellent work, and deal with obstacles in creative, successful ways, you will most likely be offered a full-time job—either there or at a similar company. Jane said that she didn't care whether you finished college, if you have already proven you are worth hiring.

She added that, for people who have not done exactly the work they are being hired to do, the most important quality is an ability to learn to do the job quickly and well. A great hire will understand what information is needed, what to do with that information, and how to deal with surprises and ambiguity.

Computers programs cannot gauge the things Bock and Swift are looking for. How could an AI program evaluate answers to interview questions when it can't even figure out what "it" refers to in a sentence? How would it quantify resilience? Data clowns assume that if you can't measure it, it must not be important. However, as sociologist William Bruce Cameron once wrote, "Not everything that can be counted counts, and not everything that counts can be counted."

Artificial intelligence programs screen job applications in simplistic and questionable ways. For example, one program concluded that, because several good programmers in its database visited a particular Japanese manga site, people who visit this site are likely to be good programmers, even though the chief scientist admitted that, "Obviously, it's not a causal relationship."

Amazon recently tried to develop customized algorithms for evaluating the resumes of applicants for hundreds of different kinds of jobs at Amazon. The algorithms trained on the resumes of job applicants over the previous ten years, and tended to favor people who were like the (mostly male) people Amazon had hired in the past. Candidates who went to all-women's colleges were downgraded because men who worked at Amazon hadn't gone to those colleges. Ditto with candidates who played on female sports teams.

Not only that, but the algorithms must have picked up coincidental patterns that had nothing to do with competency. Insiders complained that the algorithms often recommended people who had none of the requisite skills. Some said that it was as if the algorithms were flipping coins, which is what you would expect from a data-mining program that identifies patterns without being able to judge whether the patterns are meaningful.

Amazon shut down the project, with a spokesperson claiming that the algorithms were "never used by Amazon recruiters to evaluate candidates," though that claim seems contradicted by reports that insiders often found recommended candidates to be unqualified.

Amazon is hardly alone. More than half of all U.S. companies expect to use AI within the next five years to evaluate job candidates. Some will undoubtedly be creative—and baffling. Goldman Sachs and Unilever have used technology that analyzes the facial expressions and voices of job candidates while they are being interviewed. It is hard to imagine getting any farther from the sensible advice of Lazlo Bock and Jane Swift. Maybe by checking wristband sizes?

Hi-tech redlining

During the Great Depression, the federal government created the Home Owners' Loan Corporation, which made low-interest home loans, and the Federal Housing Administration (FHA), which guaranteed mortgages made by private banks. The Home Owners' Loan Corporation constructed "residential safety maps" that graded neighborhoods on a scale of A to D, with D neighborhoods color coded in red—hence the label "redlining" for undesirable, high-risk neighborhoods. These maps were also used by the FHA and private businesses, and spilled over into banking, insurance, and retail stores.

The Home Owners' Loan Corporation and the FHA were more active in California than in any other state. Security First National Bank in Los Angeles had its own neighborhood rating system. Most neighborhoods in central Los Angeles were redlined, often with explicit notations about "concentrations of Japanese and Negroes." Boyle Heights was said to be "honeycombed with diverse and subversive elements." Watts was redlined because it was a melting pot of not only Blacks and Japanese, but also Germans, Greeks, Italians, and Scots.

If you go to the MappingInequality website, you can see redlining maps for many major cities and even zoom in on specific blocks and read the disturbing comments made by the people who created these maps.

Redlining created a vicious cycle of restricted services and deteriorating neighborhoods—whose effects are still felt today. Now we are imperiled by hi-tech redlining. Employment, insurance, and loan applications are increasingly being evaluated by black-box data-mining models that are not as overt but may be even more pernicious than color-coded maps. No one, not even the programmers who write the code, know exactly how black-box

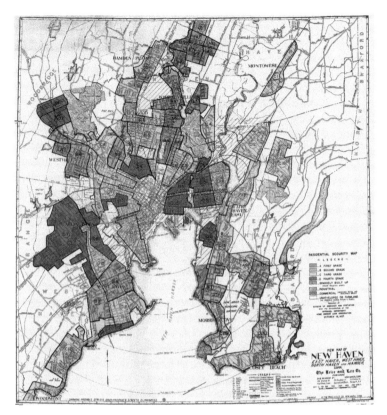

Figure 9.2 Redlining New Haven; HOLC Map of New Haven, 1937

algorithms make their predictions, but it is almost certain that employment, loan, and insurance algorithms directly or indirectly consider gender, race, ethnicity, sexual orientation, and the like. It is not moral or ethical to penalize individuals because they share characteristics of groups that a black-box algorithm has chosen as predictive of bad behavior.

Consider that algorithm that favored job applicants who visited a Japanese manga site. How fair is it if a Hispanic female does not spend time at a site that is popular with white male software engineers? How fair is it to women if they are penalized for going to women's colleges or join female organizations?

A loan-application algorithm considers how frequently incoming cell-phone calls are answered. How fair is it if certain religions are not supposed to answer the phone at certain times? An insurance algorithm is based on Facebook word choices. How fair is it if word choices are related to gender, race, ethnicity, or sexual orientation?

Two Chinese researchers recently reported that they could predict with 89.5 percent accuracy whether a person is a criminal by applying their computer algorithm to scanned facial photos. Their program found "some discriminating structural features for predicting criminality, such as lip curvature, eye inner corner distance, and the so-called nose–mouth angle." Facial recognition algorithms are inherently discriminatory.

One blogger wrote,

What if they just placed the people that look like criminals into an internment camp? What harm would that do? They would just have to stay there until they went through an extensive rehabilitation program. Even if some went that were innocent; how could this adversely affect them in the long run?

Is our faith in computers so blind that we are willing to trust algorithms to reject job applications and loan applications, set insurance rates, determine the length of prison sentences, and put people in internment camps? Favoring some individuals and mistreating others because they happen to have irrelevant characteristics selected by a mindless computer program is unconscionable hi-tech redlining.

Voter name crosschecks

Here's a more subtle example. The U.S. Interstate Voter Registration Crosscheck Program combs through a nationwide database looking for instances where two registered voters have the same first name, last name, and date of birth, suggesting voter fraud—one person voting multiple times. One study found that one in six Hispanics, one in seven Asian Americans, and one in nine African Americans were on the list. The problem is that there is less variation in names among these minorities, so they are more likely to show up with matching names. Mark Swedlund, a database expert, said, "I'm a data guy. I can't tell you what the intent was. I can only tell you what the outcome is. And the outcome is discriminatory against minorities."

OKCupid's experiments

OKCupid recently revealed three experiments they ran on their dating-site customers. In experiment 1, they temporarily removed all pictures from the site and found that there were far fewer initial messages, supporting the hypothesis that love is not blind. In experiment 2, they randomly hid people's profile text and found that it had no effect on personality ratings, supporting the hypothesis that love cannot read. These were fairly standard experiments, common across the web. Experiment 3, however, may have crossed an ethical line by turning compatibility ratings upside down. Randomly selected customers were informed that someone who was actually highly compatible with them was a bad match and vice versa. OKCupid evidently wanted to validate their compatibility ratings, but they should have considered the fact that their customers surely did not want to have their lives disrupted by romantic mismatches.

In order to help organizations navigate the ethics of experiments like these, a national commission issued the Belmont Report. It summarizes guidelines for ethical experimentation and highlights three core principles: respect, beneficence, and justice. Informed consent falls under the principle of respect. When signing up for OKCupid, people probably checked a box at the end of a long legal agreement indicating their consent for research; however, it is doubtful that this consent is "informed" in any meaningful sense. Even if they read the document, it is unlikely that they anticipated that OKCupid would use false compatibility ratings. This experiment also failed to meet the principle of beneficence. A date with an incompatible person could be excruciating; a missed date with a potential soulmate could be life changing.

Good data scientists consider the issues raised in the Belmont Report before messing around with people's lives.

Misleading graphs

Often, the most compelling way to summarize data is with a graph that is accurate and easily understood. Graphs can help us see tendencies, patterns, trends, and relationships. However, data can be distorted and mangled by graphs, either intentionally or unintentionally. Good data scientists do neither.

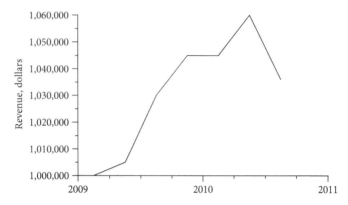

Figure 9.3 Revenue falls sharply

For example, the Chief Executive Officer (CEO) of an Internet company was perplexed and frustrated because her Board of Directors had peppered her with questions about a graph like Figure 9.3 that had been prepared by the finance department, and her analytics department hadn't alerted her to any drop in revenue.

She called a meeting to hash out the difference. When she commanded the finance people to, "Bring up the chart from the board meeting," the analytics people burst out laughing because the chart didn't have a zero on the vertical axis. The embarrassed finance guy adjusted the graph to include zero and the chart looked like Figure 9.4.

The omission of zero magnifies the ups and downs, which makes them easier to spot, but exaggerates their magnitude. The height of the line in Figure 9.3 falls by 40 percent, even though revenue only went down 2 percent.

The analytics people kept chuckling and needling the finance guy who had botched the chart, saying things like, "When revenue is trending down, please make the axis go to negative $1 million, but when revenue is going up, use the axis you used before."

The CEO was not amused. She told everyone that all future charts will have zero on the axis.

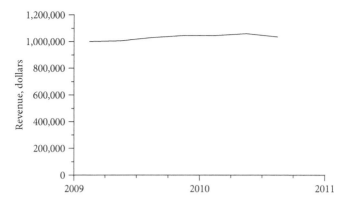

Figure 9.4 Revenue is flat

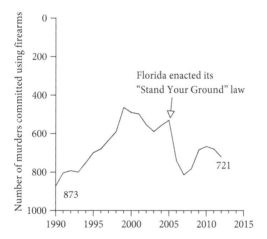

Figure 9.5 What the?

A really, really bad graph

Figure 9.5 apparently shows that murders committed with firearms in Florida declined after Florida enacted a "stand your ground" law which allows citizens to use force, without retreating, in order to defend themselves against threats or perceived threats.

Look at the vertical axis, however. It starts at 1,000 and goes backwards, up to zero. The apparent decline in murders between 2005 and 2012 was actually a 36 percent increase, from 530 to 721.

Perhaps the most perplexing thing about this topsy-turvy graph is that the person who drew it was not trying to create a false impression that murders had declined. The author was trying to demonstrate that murders had increased after the law's passage and, for some reason, thought an upside-down figure would be more persuasive!

The art of graphing

The first full-time graph specialist for *Time* magazine was an art school graduate who asserted that "the challenge is to present statistics as a visual idea rather than a tedious parade of numbers." Numbers are not inherently tedious. They can be illuminating, fascinating, even entertaining. The trouble starts when we decide that it is more important to be artistic than informative.

Here is an example. Many athletes and fans believe in the hot hand; for example, that a basketball player who makes several shots in a row is very likely to make his next shot. Three psychologists (Thomas Gilovich, Robert Vallone, and Amos Tversky) looked at a variety of basketball data and concluded that this common perception is a misperception. They compared how frequently several professional players made shots after making one, two, or three shots in a row with how frequently these players made shots after missing one, two, or three shots in a row. They found that, if anything, some players did slightly *worse* after making shots than after missing them.

There are several problems with their data. The shots might be taken 30 seconds apart, 5 minutes apart, in different halves of a game, or even in different games. In addition, a player who makes several shots may be tempted to take more difficult shots and teams may guard hot players differently, either of which would explain why players might do worse after making several shots in a row.

To get around these problems, Gary and Reid Dorsey-Palmateer (then a student, now a professor) looked at professional bowling data, a sport in which every throw is under virtually identical conditions, with little time between throws. They calculated the frequency with which professional bowlers threw strikes (knocking down all ten pins with one throw) after previously throwing strikes or non-strikes.

Figure 9.6 shows how their data might have been displayed using four pie charts, with each pie slice depicting the percent of the time that strikes and non-strikes were thrown. These 3-dimensional pies use a variety of shadings and are visually interesting, but convey little information. It is hard to gauge the relative sizes of the pie slices, and the lettering and shading are distracting. It is a visual challenge to focus on a single pie and interpret the results.

If the lettering were removed, Figure 9.6 might pass for modern art. But what does it mean? The cacophony of colors and shading is dizzying, if not nauseating, and the conclusions are elusive.

Gary and Reid skipped the pie charts and presented their results in a simple table, like Table 9.1. These professional bowlers were more likely to throw a strike after throwing strikes than after throwing non-strikes. The differences in Table 9.1 may seem modest, but they are the difference between winning and losing professional bowling tournaments.

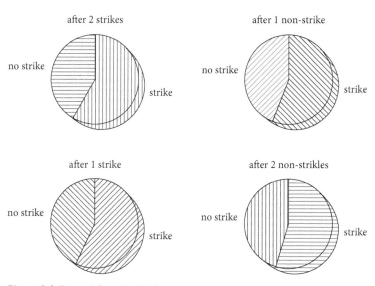

Figure 9.6 Four uninformative pie charts

Table 9.1 *Percentage strikes following strikes or non-strikes*

	Percentage strikes
After 2 strikes	58.2
After 1 strike	57.1
After 1 non-strike	56.0
After 2 non-strikes	54.6

One potential problem with the calculations in Table 9.1 is that these data aggregate different bowlers of different skills. Perhaps the reason that a strike is more likely after a strike than after a non-strike is that bowlers who throw strikes are generally better bowlers. These are all elite professional bowlers, but some are more elite than others.

So, Gary and Reid looked at each bowler individually. Was Bowler Bill more likely to throw a strike after he threw a strike, or after he threw a non-strike? They found that 60 percent of the bowlers were more likely to throw a strike after they had thrown a strike than after a non-strike, and that 70 percent of the bowlers were more likely to throw a strike after they had thrown two strikes than after two non-strikes. Bowlers do get hot (or at least warm) hands.

These 60 percent and 70 percent numbers could be shown in two pie charts, but that would not be an improvement over a simple sentence.

Effective data scientists know that they are trying to convey accurate information in an easily understood way. We have never seen a pie chart that was an improvement over a simple table. Even worse, the creative addition of pictures, colors, shading, blots, and splotches may produce *chartjunk* that confuses the reader and strains the eyes.

Fake news

See if you can spot the problems with Figures 9.7 and 9.8, which are based on charts from a "trusted news source."

Be careful what you wish for

Many people are fascinated by the idea of sending a saliva sample to a genetic-testing company and getting back a genetic-heritage profile and a

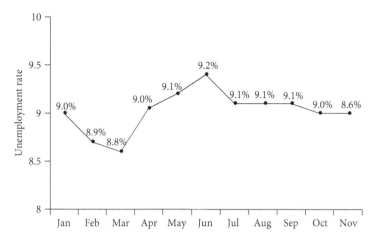

Figure 9.7 Unemployment rate under President Obama

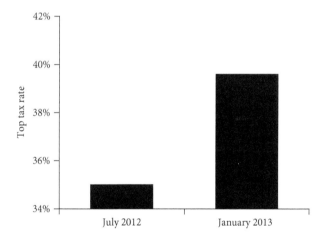

Figure 9.8 If Bush tax cuts expire

report of medical predispositions. It sounds amazing, but it can be misleading, disappointing, or worse.

Ancestry estimates are inherently volatile and often contradictory. If you send saliva samples to five different companies, you are likely to get back five different outcomes. Part of the problem is that the results depend on the database that the companies have at the time the saliva is submitted. Different companies have different databases. Even for a single company, if you send samples to the company every year for five years, the results are likely to evolve over time as the company's database grows.

Test results from a single company can vary even with little or no time between tests. In one experiment, identical triplets sent their DNA to one of the most famous labs. Even though they had exactly the same ancestors, the triplets were told that they were 11, 18, and 22 percent French or German. In another experiment, a company employee was tested every week for five weeks and got five different sets of results. When he asked a lab assistant about it, he was told to average the results—which is good for the company's revenue, not so good for a customer's confidence in the results of a single test.

An even more fundamental issue is that an ancestry DNA test compares your DNA to the DNA of people now living in different parts of the world such as Scandinavia, Southern Europe, and the Middle East. The ancestral question most people ask is, "Where did I come from? Were some of my distant ancestors Scandinavian?" We can't answer that question by looking at the current geographic distribution of DNA. Migration has made the world a genetic melting pot. Nor could we answer the ancestor question definitively by looking backward, because the answer depends on how far back we look. Five generations ago? Ten generations? A thousand years? If we look far enough back, we all may have come out of Africa (or maybe Australia).

Genetic tests can also identify our close relatives. Sometimes, we don't want to know. Families have been torn asunder by genetic tests that reveal assumed parents not to be real parents, assumed siblings not to be real siblings, and strangers to be living evidence of infidelity.

Additionally, genetic tests often do not provide reliable information about inherited conditions or disease risk. For most diseases, family history is a much better risk predictor because the gene combinations that elevate risk are complex and not well understood. Hank Greely, Director at the Center for Law and Biosciences at Stanford, argues that, "The evidence is increasingly strong that the benefits of direct-to-consumer testing for these kinds of indications are somewhere between small and zero."

Genetic testing is often surprisingly inexpensive, or we should say suspiciously inexpensive. Most testing companies don't make money on the tests; they make money selling personal information to other companies, including pharmaceutical companies that want to sell personalized health care.

Words of wisdom in the age of social media: if you're sharing information on a free service, you're not the customer, you're the product. And it is not just the person who is tested. Once a company has one person's genetic information, it also knows a lot about that person's genetic relatives, even if they never get tested. Ironically, one genetic-testing company posted this ominous warning:

Genetic Information that you share with others could be used against your interests. You should be careful about sharing your Genetic Information with others.

It sounds like a warning against other companies, but it is really a warning not to use their genetic-testing kit.

Although very few do so, an honorable company would clarify that cryptic warning and include several other disclaimers:

1 Ancestry results are uncertain and are based the current geographic dispersion of DNA.
2 Family history of disease may be better than DNA for predicting disease risks.
3 We will sell your genetic information to pharmaceutical companies.
4 Genetic testing can be paternity testing, which sometimes ends in tears.

We should mention that although sharing DNA is currently unlikely to help identify past ancestors or future diseases, one group does benefit from the data collection: law enforcement agencies. Genetic genealogy has led to arrests for at least a dozen unsolved violent crimes. For example, the Golden State Killer committed dozens of rapes and burglaries and at least 13 murders in California between 1974 to 1986. The police had his DNA for 40 years, but no leads. After investigators uploaded the killer's DNA to a popular genealogy site, they found relatives in the database and were able to determine that the DNA came from Joseph James DeAngelo, a retired police officer. According to 60 Minutes producer Michael Karzis, this new use of DNA databases is here to stay: "The horses have left the stable. I don't think they're going to go back in."

Unintended consequences

Actions have consequences, but not always the consequences that were intended. Sometimes the unintended consequences are positive; for example, sunken ships can create coral reefs. More often, the intention is good, but the unexpected consequences are bad.

Australia is an isolated continent with more than 80 percent of its plants, mammals, and reptiles unique to Australia—including kangaroos, koalas, and wallabies. The well-meaning introduction of species from other continents has not always gone well. When England sent 11 ships (the First Fleet) to start a penal colony in Australia, they brought along rabbits to be bred as food. The rabbits "bred and bred and bred" and have had devastating effects on Australia's ecology.

In the 1930s, cane toads were brought to Australia to feast on the cane beetles that were attacking Australia's sugar cane fields. Like the rabbits before them, cane toads are prolific breeders and have become a major pest. (They were also ineffective in controlling sugar cane beetles; the beetles like living at the top of the sugar cane and toad canes can't jump very high.) Toad canes are large (4 to 6 inches long, weighing about 6 pounds) and ugly and their poisonous glands can be deadly for lizards, snakes, and crocodiles that eat them. They now number more than 1.5 billion; that's 75 cane toads for every Australian. They are also lazy and like to travel on paved roads, especially at night—which creates unwanted hazards for some cars and a ghoulish sport for others.

Good data scientists not only think about their goals, but also about the possible unintended consequences of their actions.

I've been hacked!

An Internet company (DomainsAreUs) had purchased hundreds of thousands of previously owned websites with the intention of showing ads to people who came to these recycled websites. For example, a person who bookmarked a site that once sold baseball cards might go to the site looking for baseball cards and find advertisements for products that DomainsAreUs thought would appeal to baseball-card enthusiasts.

One day, some data scientists at DomainsAreUs noticed that most of the "visits" to their recycled sites were automated requests for images that used to be displayed on the websites. As an example, suppose someone wanted to use a picture of Mickey Mantle, a famous New York Yankees baseball player, as an avatar on a baseball forum, and had provided a hyperlink to a picture of Mickey Mantle on a baseball-card website. Every time this person posted comments on the forum, the Mickey Mantle image would be sent from the baseball website to the forum.

Unfortunately, when the baseball-card company let its rights to the site expire and DomainsAreUs picked it up, all the previous content disappeared, including the Mickey Mantle image. So, instead of a picture of Mickey Mantle, the forum avatar became an error icon signaling a broken link.

DomainsAreUs was often creative in finding value in their data and they realized that they could respond to those image requests with anything they wanted. The tech team ran an experiment. They created a cryptic domain name like WQ534X.bit.ly and an "image" that simply said "WQ534X.bit.ly" in response to requests for deleted images. Then they counted how often people typed WQ534X.bit.ly into their browser and visited the site out of curiosity.

The analytics team heard about this experiment and thought it was pointless. Instead of a meaningless domain name, why not send a recognizable logo for one of DomainsAreUs's real websites? Free publicity! Potentially millions of people would see the logo and find their way to the DomainsAreUs website and perhaps buy something. They selected the image for a DomainsAreUs financial-advice website and waited for customers.

The analytics team initially thought it was a great idea. They found the logo everywhere: forums, home pages, and, of course, avatars. Some pages had a giant copy of the finance logo across the top (DomainsAreUs didn't control the size of the image) and several copies of the logo wallpapering the rest of the page.

What the DomainsAreUs analytics team failed to consider was how people would react to this intrusive advertising. Users were nervous and angry because they thought that a nefarious finance site had hacked their computers, replaced their avatars, and might be stealing personal information. Word quickly spread that the DomainsAreUs finance site was hacking computers!

The upper management was not amused. It is sometimes said that there is no such thing as bad publicity, but this was very bad publicity that could give their finance site a terrible reputation. They killed the experiment, and everyone who was not directly involved pretended they had nothing to do with it. The moral of the story is to consider unintended consequences, even if it's "just an experiment."

The Streisand effect

The Streisand effect is named after singer Barbra Streisand, who sued a website for posting a photo of her lavish beachfront home. Before her suit was filed, only six people had seen the photo—and two of them were her attorneys. After the suit became public, hundreds of thousands of people went to the site to see the photo.

The Streisand effect cuts two ways. People who don't want publicity should be wary of censorship attempts that bring publicity. On the other hand, people who do want publicity can get it by provoking someone to try to suppress them.

Perverse incentives are a variation on the phenomenon of unintended consequences. For example, the British colonial government in India tried to reduce the number of venomous cobras in Delhi by paying bounties to people who brought in dead cobras. The hope was that this incentive would encourage people to kill cobras. Guess what happened. Enterprising citizens bred cobras that they could kill and cash in. When the authorities learned of this scheme, they ended the bounties, and the breeders freed their cobras—resulting in a cobra population much larger than before the program began.

Whenever bosses set goals carelessly, you can bet that perverse incentives have been created. An employee at one Internet company was offered a bonus if he could hit a target number for revenue per visitor. It was a sensible goal, but there was an easy, unethical way to reach the target. U.S. web traffic generates more revenue per visitor than international traffic, so the employee blocked international traffic from the company's web pages. Overall revenue dropped, but revenue per visitor went through the roof. Mission accomplished.

Jay was once given a surprising goal...

"Your team needs to buy 100,000 domain names this month."
"Don't you mean 100,000 profitable names?"
"No, we're just focused on revenue right now."

"Well that's easy, we can buy 100,000 domain names tomorrow."
(silence)
"You know what I mean."

In the same way, computer algorithms intended to identify or assess people's behavior may encourage people to change their behavior in ways that make the algorithms useless or perverse.

Our fundamental right to privacy

In 1961, at the end of his eight years as U.S. President, Dwight D. Eisenhower gave a nationally televised farewell speech in which he warned the nation of the growing power of an alliance of the armed forces and the defense contractors that he called "the military–industrial complex." Military leaders and defense contractors campaign for a growing arsenal of expensive weapons that makes the military stronger and contractors richer. Eisenhower's speech was particularly striking because Eisenhower had been a five-star general and the Supreme Commander of the Allied forces fighting Germany in northwest Europe in World War II.

Not only can a strong military seduce governments into launching unjustified and ill-conceived military adventures, such wars may actually benefit the military–industrial complex. When weapons are destroyed, the military can seek funding for new and improved weapons, which industry is happy to supply. Without continuing conflicts, the military has less justification for weapon requests, and defense contractors have fewer weapon orders. "Wars for profit" not only distort foreign policy, but also undermine freedom and well-being as citizens are sent off to fight wars while resources are diverted from building homes and schools to supplying bombs and tanks.

In 2018, Apple CEO Tim Cook invoked memories of Eisenhower when he warned of the growing power of the "data–industrial complex." This alliance is between those who collect and analyze personal data and the businesses and governments that use these personal data. Data can be used not only as "weapons" against foreign governments, but against a country's own citizens.

Businesses use data to manipulate our spending. The costs are not only the money squandered on things we don't need, but, more profoundly, a loss of privacy. There is no compelling reason for businesses to know our income, hobbies, and sexual preferences.

Politicians use data to manipulate our votes. The costs are not only the distortion of our opinions, but, again, a loss of privacy. There is no compelling reason for politicians to know our personal views about them and their positions. There are good reasons for secret ballots.

Governments use data to monitor us. The costs are not only the resources used to track what we see and do, but, once again, a loss of privacy. There is no compelling reason for governments to know who our friends are, what we read and write, where we go, and what we do.

Cook argued that,

Our own information—from the everyday to the deeply personal—is being weaponized against us with military efficiency.... Every day billions of dollars change hands and countless decisions are made on the basis of our likes and dislikes, our friends and families, our relationships and conversations, our wishes and fears, our hopes and dreams.... We shouldn't sugarcoat the consequences. This is surveillance....For artificial intelligence to be truly smart it must respect human values—including privacy. If we get this wrong, the dangers are profound.

Cook proposed four key components of a digital privacy law:

Minimization.	Nonessential data should either not identify individuals or not be collected at all.
Transparency.	Individuals should know what data are being collected and why.
Access.	Our data belong to us and it should be easy to see, correct, and delete personal data.
Security.	Data should be stored securely to safeguard privacy.

Freedom of expression

Repressive governments know that privacy and freedom of expression are critical for an informed citizenry to know what government officials are doing and to hold governments accountable for their actions. Citizens who read what they want, think what they want, write what they want, and say what they want are threats to totalitarian governments.

Governments do not need mass incarcerations to suppress their citizens. People will self-censor if they know that it is likely that they are being monitored and that they may possibly be punished. The mere existence (or even rumors of the existence) of mass surveillance programs can crush freedoms.

After Edward Snowden revealed that the National Security Agency (NSA) was using Facebook, Google, and Microsoft servers to track online communications, Wikipedia searches for keywords like *jihad* and *chemical weapon* declined precipitously, presumably because of people's fears that the NSA or some other government agency might be monitoring them. Better to avoid suspicion than to risk the wrath of government bureaucrats.

Snowden also revealed that a secret court order had compelled Verizon to give the NSA all of its telephone records on an "ongoing daily basis." People who assume that the NSA is listening to their phone conversations may understandably avoid using words that might be misconstrued. When citizens assume that their government is surveilling every action, it is like living in a prison.

In a contest for the most repressive country on earth, North Korea is a strong medal contender. The government's propaganda machine endlessly promotes a cult of personality surrounding Kim Jong-un, the 3rd Supreme Leader of North Korea (after his father and grandfather). The Kim family is touted to be god-like in that it is infallible and it is a crime to question or criticize it. The government punishes criticism or resistance, even from the relatives of top government officials, with prison or death.

According to the government propaganda machine, North Koreans enjoy one of the world's highest standards of living, for which they must be eternally grateful for the wise guidance of their Supreme Leader. The reality is that its per capita GDP is around $1,300, with much of it devoted to the military and maintaining the lifestyle of the ruling class. Most workers are paid less than $3 a month and it has been estimated that 40 percent of the population are undernourished. The World Food Programme estimated that 28 percent of North Koreans have stunted growth. The National Intelligence Council estimated that 25 percent of North Korean military conscripts are disqualified because of mental disabilities, most likely related to poor nutrition.

To maintain this fiction of North Korea's high standard of living, the government relentlessly suppresses contact with the outside world, including foreign newspapers, magazines, radio and TV stations, and the Internet. For example, anyone caught having a radio capable of receiving foreign stations can be sent to a prison camp.

Several refugee organizations have been trying to undermine the North Korean regime simply by showing citizens glimpses of the world outside North Korea. Tens of thousands of USB drives that can be played on cheap Chinese-made USB-compatible video players have been smuggled into North Korea by balloons and drones or by bribing guards on the Chinese border.

What is on these forbidden USB drives? Wikipedia downloads and movies, music, and TV shows, like *Friends*, which show what life is like outside North Korea. One activist, argued that, "They see the leisure, the freedom. They realize that this isn't the enemy; it's what they want for themselves. It cancels out everything they've been told. And when that happens, it starts a revolution in their mind."

Not surprisingly, the North Korean government has fought back, arresting and executing people caught with "illegal media." Kim Jong-un vented that North Korea needed to "take the initiative in launching operations to make the imperialist moves for ideological and cultural infiltration end in smoke" and to set up "mosquito nets with two or three layers to prevent capitalist ideology, which the enemy is persistently attempting to spread, from infiltrating across our border."

Computers can be used for good, which is why evil governments don't like them. Computers can also be used for evil.

The Golden Rule

An unfortunate reality in the age of big data is Big Brother monitoring us incessantly. Big Brother is indeed watching, but it is big business as well as big government collecting detailed information about everything we do so that they can predict our actions and manipulate our behavior. Big business and big government monitor our credit cards, checking accounts, computers, and telephones, watch us on surveillance cameras, and purchase data from firms dedicated to finding out everything they can about each and every one of us.

Good data scientists proceed cautiously, respectful of our rights and our privacy. The Golden Rule applies to data science: treat others as you would like to be treated.

The Ninth Pitfall of Data Science is:

Doing Harm

The Great Recession

> Ratings agencies continue to create an even bigger monster—the
> CDO market. Let's hope we are all wealthy and retired by the time
> this house of cards falters.

This ironic email was written by a Standard & Poor's employee who was worried about the high ratings they were giving to dodgy mortgage-related securities. If there was a ground zero for the financial crisis that triggered the Great Recession that began in 2007, it was triple-A ratings for securities that turned out to be worthless. A deeper problem was that data clowns put enough rocket fuel into the financial system to blow it sky high. Let's revisit the disaster and the role of Data Science Pitfalls.

Decades ago, people bought homes by getting mortgages from local banks—with good reason. New York banks didn't want to lend money to a family they never met that wanted to buy a home in a California city called Rancho Cucamonga that the bank wasn't sure really existed.

Then, in the 1970s, the federal government revolutionized mortgage lending by creating three government sponsored agencies, nicknamed Fannie Mae, Ginnie Mae, and Freddie Mac, that raised money from investors in order to buy mortgages, the same way that stock mutual funds raise money from investors in order to buy stocks. Soon private mortgage funds joined the revolution, and more than half of all mortgages were sold to public and private mortgage funds. Everyone could invest in mortgages, and banks that sold their mortgages got more money to make more loans. Win. Win.

Unfortunately, this new system created perverse incentives. Banks had traditionally been cautious lenders, because if a borrower defaulted, the bank was stuck with the bad loan. However, when a bank sells its mortgages to someone else it has little reason to be prudent, because the bank's profits now depend on the number of loans it approves, not on whether the monthly mortgage payments are made on time.

Once private mortgage funds with lower standards entered the game, banks could increasingly make loans to "subprime" borrowers with low credit ratings and modest income. This problem was exacerbated by the growth of national mortgage brokers who seemed to lend to anyone, anywhere. They made money by making mortgages, and making mortgages was what they did, including NINJA loans (applicants with No Income, No Job, and no Assets who, at the first sign of trouble, disappear like ninjas in the night).

Too many loans were made (often at initially low teaser rates) to people who had no realistic prospects of making the mortgage payments month after month. The website of one mortgage broker boasted, "We don't get paid unless we say YES." With no incentive to say NO, it is not surprising that lots of questionable loans were approved.

Who was dumb enough to invest in these risky pools of subprime mortgages? Incredibly, major buyers included pension funds, insurance companies, and other institutions that are supposed to stick to ultrasafe triple-A rated securities. Conveniently enough, mortgage pools of subprime loans were rated triple-A by credit-rating agencies so that supposedly prudent investors could buy them. How could a pool of mortgages be rated triple-A when almost every mortgage in the pool was junk?

There is a simple, but elegant, theorem that states that a bundle of a risky assets is safer than any asset in the bundle. This theorem is the basis for a famous mathematical paper by Nobel Laureate Paul Samuelson titled, "General Proof that Diversification Pays." Here is one of his theorems:

Theorem 1: If $U(X)$ is a strictly concave and smooth function that is monotonic for non-negative X, and $(X_1,...,X_n)$ are independently, identically distributed variates with joint frequency distribution

$$\text{Prob}\{X_1 \leq x_1,...,X_n \leq x_n\} = F(x_1)F(x_2)...F(x_n)$$

with $E[x_i] = \int\limits_{-\infty}^{\infty} X_i\, dF(X_i) = \mu_1$

$E[x_i - \mu_i]^2 = \int\limits_{-\infty}^{\infty} (X_i - \mu_1)^2\, dF(X_i) = \mu_2$

with $0 < \mu_2 < \infty$,

then

$$E\left[U\left(\sum_1^n \lambda_j x_j \right) \right] = \int\limits_{-\infty}^{\infty} \cdots \int\limits_{-\infty}^{\infty} u\left(\sum_1^n \lambda_j x_j \right) dF(x_1)\ldots dF(x_n)$$

$$= \psi(\lambda_1,\ldots,\lambda_n)$$

is a strictly concave symmetric function that attains its unique maximum, subject to $\lambda_1 + \lambda_2 + \ldots + \lambda_n = 1$, $\lambda_i \geq 0$

at $\psi\left(\dfrac{1}{n},\ldots,\dfrac{1}{n} \right)$.

That is some pretty serious math, so it must be right. Right? Well, sort of.

The crucial assumption, as Samuelson recognized, is that the returns are independent. In 100 coin flips, the average outcome (close to 50 percent heads) is more certain than is the outcome of any single flip. Diversification pays. Not so, if we buy 100 shares of Under Armour stock. If one share does poorly, so will the other 99 shares.

A pool of 100 mortgages is in between these extremes of 100 coin flips and 100 shares of the same stock. If Jane loses her job and can't make her mortgage payments, that doesn't necessarily mean that Joe will lose his job and not make his mortgage payments. However, in a recession, a lot of Janes and Joes lose their jobs, especially Janes and Joes who were bad credit risks because of their unstable employment history.

In addition, many NINJA borrowers were counting on home prices rising so that they could sell their homes before their teaser loan rates expired. When home prices stopped rising and started falling, many NINJA borrowers were in trouble. They couldn't make the higher mortgage payments and their mortgages were "underwater," in that they couldn't sell their homes for enough money to pay off their mortgages.

Many walked away, abandoning their homes and leaving mortgage pools with underwater mortgages on depreciating homes.

Too many blithely believed that diversification makes a pool of junk mortgages safe:

Some clever financial engineers discovered that they could make more fees by dividing mortgage pools into sub-pools, called *tranches*, based on the default risk. Investors who bought the lowest ("junior") tranches received more income from the mortgages, but were hit the hardest if borrowers defaulted. These sliced-and-diced mortgage pools were called *collateralized debt obligations* (CDOs).

As if that wasn't complicated enough, investment banks created *CDOs-squared*, which are CDOs made up of the tranches of other CDOs. A junior CDO-squared consists of junior tranches from several CDOs. In theory, this offered more diversification than a junior tranche from a single CDO. This was, again, the diversification fallacy that applied the mathematical analysis of independent investments to investments that are not independent. If falling home prices caused one junior CDO to crash, many would.

Derivatives

A financial derivative is a security whose value is derived from another asset. For example, the value of a call option that gives the owner the right to buy Apple stock at a specified price depends on the market price of Apple stock.

One thing that so-called financial engineers are very good at is thinking up exotic derivatives that sound good, but almost no one understands. The legal documents describing derivatives might be thousands of pages long; the sales pitch might last 30 minutes. Investors who were persuaded by the slick pitch ignored the sound advice, "Don't invest in something you don't understand." They did not avoid

In addition to CDOs and CDO-squareds, financial engineers created a derivative called a *credit default swap* (CDS). The buyer of a credit default swap pays a fee to pass the default risk on to the seller. For example, a pension fund might own a junior tranche of a CDO, and be concerned about loan defaults. The pension fund pays a fee to a third party that agrees to buy the junior tranche at a fixed price if there are defaults. The pension fund has effectively bought default insurance.

Institutions that sell credit default swaps are essentially insurance companies and should set the CDS terms based on their assessment of default risks. They might believe that a diversified portfolio of credit default swaps is safe. They will be in a world of trouble, however, if a macroeconomic event, like a global recession, causes a wave of defaults—just like insurance companies operating in Chicago in the nineteenth century that thought they were financially invulnerable because they insured thousands of homes, but were bankrupted by the Great Chicago Fire of 1871.

Many investors didn't buy credit default swaps for insurance, but for speculation. A hedge fund might pay $1 million a year for a credit default swap on a $100 million subprime CDO that it doesn't own. The hedge fund was not being prudent by buying insurance; it was making a 100-to-1 bet on a mortgage meltdown. If the CDO doesn't default, the hedge fund is out $1 million. If the CDO does default, the hedge fund makes $100 million.

If that wasn't enough, financial engineers created synthetic CDOs which are essentially wagers on CDSs, which are wagers on real CDOs! Confusing? Yes, but that didn't seem to matter. The appealing thing about synthetic CDOs is that they were not limited by the number of mortgages or CDOs. Speculators could make 1,000 bets on one CDO defaulting. Fees were multiplied 1,000-fold, and so were the financial consequences of a default. Imagine the damage to the financial system that the Great Chicago Fire could have caused if insurance companies all over the country had made thousands of wagers that there would not be a major fire in Chicago.

Some investment banks were double dipping, generating fees from selling CDOs while simultaneously using credit default swaps and synthetic CDOs to bet that the CDOs they had created would default. It was

Pitfall 9: Doing Harm

Yes, CDOs, CDO-squareds, CDSs, and synthetic CDOs are baffling, bordering on incomprehensible. That is why Warren Buffett and other sensible investors buy stocks they understand: Coca-Cola, Benjamin Moore paint, or See's Candies. When financial institutions trade derivatives that they barely understand and cannot reasonably value, we have what Buffett called "financial weapons of mass destruction."

Some suspected that the risks were real, and enormous, but they didn't want to miss out on the money that was being made. Shortly before the party ended, Citigroup's CEO said that "as long as the music is playing, you've got to get up and dance. We're still dancing."

Modeling risk poorly

It is very difficult to value mortgages. How do you estimate the chances that a borrower you don't know in a city you've never heard of will default on a mortgage? How do you estimate the chances that a borrower will change jobs, marry, divorce, have children, or have some other reason for paying off a mortgage early? (Even more difficult to value are pools that include mortgages, car loans, credit card debt, and other loans.)

If we can't value the mortgages in a pool, how are we supposed to value a CDO, CDO-squared, CDS, or synthetic CDO? Professors at prestigious business schools and financial engineers with advanced degrees from top schools built mathematical models to assess the risks, but their models were deeply flawed.

One problem was that the models often focused on fluctuations in the market prices of investments, rather than the chances of a catastrophic meltdown. Citigroup's chief financial officer later said that they underestimated the risk because, "We had a market-risk lens looking at those products, not the credit-risk lens." By thinking that the relevant question was the day-to-day volatility in the market prices of CDOs, rather than the chances that mortgage defaults would make CDOs worthless, they did not avoid

Pitfall 6: Fooling Yourself

Another problem was that the models did not consider the fact that derivative contracts often require institutions that are losing money to put up more collateral. For example, if the market price of a credit default swap issued by AIG rises because the risk of default is increasing, AIG must put up more collateral to demonstrate that it can make the payout if needed. Collateral requirements magnified the financial strain on AIG, but the models ignored them, either because the model builders didn't know about them or because they are difficult to model. Either way, they did not avoid

Yet another problem was that the model builders were often seduced by the mathematical appeal of the normal distribution. The models ignored fat tails and underestimated the chances of extreme events. A related problem was that the models often used the convenient assumption that mortgage defaults are independent. By not adequately considering the possibility that a macroeconomic event like an economic recession would cause an avalanche of mortgage defaults, they did not avoid

Pitfall 3: Worshipping Math

Another problem was that the probability that a mortgage might default was estimated from historical data. That seems reasonable, except, historically, subprime borrowers had not gotten mortgages, so there were no data on how often they defaulted, let alone the chances they would default in a macroeconomic crisis. AIG claimed that, based on historical data, the chances that it would have to make credit default swap payouts were "remote, even in severe recessionary market scenarios." In 2007, they boasted that, "It is hard for us, without being flippant, to even see a scenario within any kind of realm of reason that would see us losing one dollar in any of those transactions." Like those who thought black swans were impossible, they considered credit default swaps risk-free because they had never seen a default:

Pitfall 1: Using Bad Data

A senior executive at AIG said that one reason they sold too many credit default swaps was that they trusted a computer model created by a business school professor. The model said that the premiums paid to AIG were free money because there was no chance that they would ever have to pay anything to cover defaults. They trusted a computer model that they did not understand:

Pitfall 4: Worshipping Computers

They trusted a computer model because it was developed by a business school professor:

Pitfall 6: Fooling Yourself

Ratings laundering

At the time, only six private companies in the United States were rated triple-A, so how could trillions of dollars in risky mortgage pools be rated triple-A? If it hadn't been for these unrealistic ratings, investors would not have trusted mortgage pools. Someone was gaming the system, big time.

A conflict of interest was created by the fact that ratings agencies were paid by the clients who ran the mortgage pools, not by the investors they are supposed to protect. Rating agencies that gave generous ratings got more business and more fees—so they gave generous ratings.

In addition, if the initial rating was low, investment banks shuffled the mortgages in their pools and sent the new pools back to the rating agencies again and again until they got triple-A ratings. This repackaging is analogous to p-hacking and therefore falls into

Pitfall 5: Torturing Data

Overall, U.S. Securities and Exchange Commission (SEC) commissioner Kathleen Casey stated it plainly: the ratings were "catastrophically misleading," but the credit rating agencies "enjoyed their most profitable years ever" while handing them out. It was

Pitfall 9: Doing Harm

The requirements for triple-A ratings are supposed to be secret so that investment banks can't game the system. However, there is an easy workaround: hire analysts who work at the rating agencies and know the secret details.

For example, a triple-A rating requires an average credit score of about 615, but not all averages are the same. A pool of mortgages with uniform 615 ratings has less risk than does a pool of mortgages with 550 ratings for half the mortgages and 680 ratings for the other half. Once Wall Street figured out this loophole, they were able to secure triple-A ratings by pairing low-risk mortgages with high-risk mortgages instead of separating the mortgages into high-risk and low-risk pools.

Another trick was to persuade immigrants to buy homes. Since they had never borrowed before, they had deceptively high credit scores. Newspaper headlines were soon reporting "$14,000 per-year field worker buys $720,000 home," which encouraged more subprime borrowers to apply for mortgages to buy homes they couldn't afford. Too many lenders were too happy to oblige:

These weren't the only blind spots in the rating agencies' models. Some didn't consider whether a loan had sufficient documentation or whether the down payment was large enough. NINJA loans were fine:

Too big to fail?

The U.S. Treasury and Federal Reserve have often operated under the assumption that some companies are "too big to fail," meaning that because of the interconnected nature of the financial system, a large company's failure might have ripple effects that could cripple the economy. If a large bank fails, it won't be able to repay the money it owes other banks, which may make it impossible for these other banks to repay their debts.

Perhaps even worse, if companies are worried about a company failing, they won't lend them money or sell them things for fear that they won't be repaid. For a bank, this might mean that they can't get money to make loans; for an industrial company, this might mean that they can't borrow money to buy the raw materials they need to keep operating. It becomes a self-fulfilling prophecy in that, unable to borrow money, a company that is feared to be in danger of failing does fail. The larger the company, the larger the risk of a contagious collapse. Thus some companies are considered too big to fail. If endangered, the Treasury and Fed step in to lend it money to save the company—and the economy.

The unintended consequence is that companies that consider themselves too big to fail may take excessive risks, knowing that they will reap outsize gains if their gambles pay off and that they will be bailed out by the government if their gambles fail.

You already know how this ended. Home prices stopped rising and the dominos started falling. Subprime borrowers couldn't make their mortgage payments and couldn't sell their homes for enough money to pay off their mortgages. Many sold their homes at distress prices or walked away from their underwater loans and let the banks sell their homes at distress prices, which put downward pressure on home prices and encouraged more subprime borrowers to walk away.

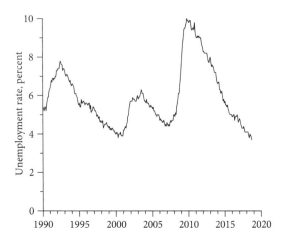

Home builders stopped building homes and ripples from unemployed construction workers spread through the economy, as spending, income, and stock prices collapsed. Household wealth fell by $12 trillion (17 percent), and employment dropped by 9 million. The graph shows that the unemployment rate doubled in 18 months, from 5 percent in April 2008 to 10 percent in October 2009, the highest since the Great Depression in the 1930s, and the U.S. economy was on the brink of a second great depression.

The mortgage defaults triggered CDO defaults, which triggered credit default swap payments. Bear Stearns, Lehman Brothers, Merrill Lynch, J.P. Morgan, AIG, and other financial institutions that had bought CDOs or sold CDSs (or both) were in trouble—and so were the financial institutions that were counting on being paid by them.

In March of 2008, Bear Sterns was rocked by losses on its subprime mortgages, mortgage-backed securities, and CDOs. The Federal Reserve acted quickly to supply $29 billion to finance J.P. Morgan Chase's purchase of Bear Stearns, explaining that it wanted to avoid a "chaotic unwinding" of Bear Sterns' investments.

Six months later, Lehman Brothers was hit by massive losses from its subprime mortgage securities. Lehman had created a lot of CDOs and ended up holding many of the junior tranches. We don't know if this was because Lehman couldn't offload the junior tranches to other investors or if Lehman thought the junior tranches were a profitable investment. We do know that Lehman was speculating with billions of borrowed dollars. When the profits on their investments were larger than the interest on their loans, Lehman's employees and stockholders made a lot of money. Lehman's CEO, Richard Fuld, was paid $485 million between 2000 and 2007. Lehman was also heavily involved in credit default swaps. Ironically, $400 billion in swaps contracts had been created by investors who wanted to bet for and against Lehman's bankruptcy.

At the same time that the federal government was pushing Bank of America to acquire financially troubled Merrill Lynch, the Treasury and Fed stood aside and let Lehman fail.

Ironically, Lehman's failure confirmed the too-big-to-fail argument by precipitating a financial crisis and near meltdown of financial firms and markets worldwide. Congress authorized a $700 billion bailout fund, of which $160 billion was used to prop up AIG and another $170 billion was used to stabilize the nine largest U.S. banks. Citigroup received $45 billion; J.P. Morgan and Wells Fargo both received $25 billion; Goldman Sachs got $10 billion.

The bailout and the Federal Reserve's aggressive monetary policy avoided a second great depression, but it took five years of economic misery to get the unemployment rate back down to 5 percent.

Andrew Redleaf, a man who ran a hedge fund called Whitebox Advisors, said that Wall Street had "substituted elaborate, statistically based insurance schemes that, with [a belief in] efficient financial markets, were assumed to make old-fashioned credit analysis and human judgment irrelevant."

Underlying many of the mistakes was incredible hubris. Too many people thought that, because they were making so much money, they must be really smart—too smart to make mistakes:

Pitfall 6: Fooling Yourself

Hundreds fell into Pitfalls. Millions paid the price.

BIBLIOGRAPHY

Abramoff, M.D. and Lou, Y., et al. 2016. Improved automated detection of diabetic retinopathy on a publicly available dataset through integration of deep learning, *Investigative Ophthalmology & Visual Science*, 57 (13), 5200–6.

AdExchanger. 2017. The top ten programmatic advertisers, *AdExchanger*. March 20. Accessed November 3, 2018. https://adexchanger.com/advertiser/top-ten-programmatic-advertisers/.

Alba, Davey. 2011. How Siri responds to questions about women's health, sex, and drugs, *Laptop*, December 2.

Alexander, Harriet. 2013. "Killer Robots" could be outlawed, *The Telegraph*, November 14.

Allan, Nicole and Thompson, Derek. 2013. The myth of the student-loan crisis, *The Atlantic*, March.

American Cancer Society Medical and Editorial Content Team. 2018. Aspartame, *American Cancer Society*, April 30. Accessed November 3, 2018. https://www.cancer.org/cancer/cancer-causes/aspartame.html.

Anderson, Chris. 2008. The end of theory, will the data deluge make the scientific method obsolete?, *Wired*, June 23.

Angwin, Julia and Larson, Jeff, et al. 2016. Machine bias, *ProPublica*, May 23. Available from: https://www.propublica.org/article/machine-bias-risk-assessments-in-criminal-sentencing.

Angwin, J., Scheiber, N. and Tobin, A. 2017. Machine bias: dozens of companies are using Facebook to exclude older workers from job ads, *ProPublica*, December 20. Available from: https://www.propublica.org/article/facebook-ads-age-discrimination-targeting.

Anonymous. undated. 8 longest 7-man checkmates, *Articles. Lomonosov Tablebases*. Accessed November 3, 2018. http://tb7.chessok.com/articles/Top8DTM_eng.

Anonymous. undated. General terms and conditions, *Full Tilt*. Accessed November 2, 2018. https://www.fulltilt.com/poker/promotions/terms/.

Anonymous. undated. Open practice badges, *Association for Psychological Science*. Accessed November 3, 2018. https://www.psychologicalscience.org/publications/badges.

Anonymous, undated. Electric & magnetic fields. National Institute of Environmental Health Sciences. Accessed November 22, 2018. https://www.niehs.nih.gov/health/topics/agents/emf/index.cfm.

Anonymous. 2007. Yearly income, $14,000. Purchase of house, $720,000. Have we all lost our minds???, *Dr Housing Bubble*, May 3. Accessed November 2, 2018. http://www.doctorhousingbubble.com/yearly-income-14000-purchase-of-house-720000-have-we-all-lost-our-minds/.

Anonymous. 2015. A long way from dismal: economics evolves, *The Economist*, 414 (8920), 8.

Anonymous. 2018. The clouds of unknowing, *The Economist*. March 18. Accessed November 19, 2018. https://www.economist.com/briefing/2010/03/18/the-clouds-of-unknowing.

Apple, Sam. 2017. The young billionaire behind the war on bad science, *Wired*. June 3.

Anonymous. 2018. DeepMind's Go playing software can now beat you at two more games. *New Scientist*. Accessed December 10, 2018. https://www.newscientist.com/article/2187599-deep-minds-go-playing-software-can-now-beat-you-at-two-more-games.

Baard. 2009. Brian Hastings wins $4M from Isildurl, *High Stakes Database*. December 9.

Bagai, Jeremy Paul. 2001. *Classic Backgammon Revisited: 120 Previously Published Problems – Reanalyzed*. Portland, OR: Flaming Sparrow Press.

Balof, Barry A. et al. 2019. Bingo paradoxes. In *The Mathematics of Various Entertaining Subjects, Vol. 3: The Magic of Mathematics*. Princeton, NJ: Princeton University Press.

Baskin, Ben. 2015. Best year ever, *Sports Illustrated*, August 24.

Belluz, Julia. 2015. John Ioannidis has dedicated his life to quantifying how science is broken, *Vox*. February 16.

Belluz, Julia. 2018. Is your cellphone giving you cancer? A comprehensive guide to the messy, frustrating research, *Vox*, July 18. https://www.vox.com/2018/7/16/17067214/cell-phone-cancer-5g-evidence-studies.

Bem, D.J. 2011. Feeling the future: experimental evidence for anomalous retroactive influences on cognition and affect, *Journal of Personality and Social Psychology*, 100 (3), 407–25.

Benjamin, Arthur, Kisenwether, Joseph and Weiss, Ben. 2007. "The BINGO Paradox." *Math Horizons* 25, 1: 18–21. doi:10.4169/mathhorizons.25.1.18.

Benjamin, Arthur T. and Quinn, Jennifer J. 2003. *Proofs That Really Count: The Art of Combinatorial Proof*. Washington, DC: Mathematical Association of America.

Berk, R. 2013. Algorithmic criminology, *Security Informatics*, 2: 5. https://doi.org/10.1186/2190-8532-2-5.

Berlau, John. 2013. Pressure builds on AIG as borrowing spree grows, *Business Insurance*, October 13, 31.

Bias, Randy. 2016. The history of pets vs cattle and how to use the analogy properly, *Cloudscaling*, September 29.

Bilanuik, Stefan. 1991. Is mathematics a science? Accessed November 2, 2018. http://euclid.trentu.ca/math/sb/misc/mathsci.html.

Boggs, Will. 2018, Researchers often ask statisticians for inappropriate analyses, *Reuters*, October 11.

Bolen, J., Mao, H. and Zeng, X. 2011. Twitter mood predicts the stock market, *Journal of Computational Science*, 2 (1), 1–8.

Brodski, A. and Paasch, G.F., et al. 2015. The faces of predictive coding, *Journal of Neuroscience*, 35 (24): 8997.

Brown, Kristen V. 2018. How DNA testing botched my family's heritage, and probably yours, too, *Gizmodo*, January 16.

Brown, N.J.L. and Coyne, J.C. 2018. Does Twitter language reliably predict heart disease? A commentary on Eichstaedt, et al. (2015a), *PeerJ* 6:e5656. https://doi.org/10.7717/peerj.5656.

Calude, Cristian S. and Longo, Giuseppe. 2016. The deluge of spurious correlations in big data, *Foundations of Science*, 22, 595–612. https://doi.org/10.1007/s10699-016-9489-4.

Camerer, Colin F. and Dreber, Anna, et al. 2018. Evaluating the replicability of social science experiments in *Nature* and *Science* between 2010 and 2015, *Nature Human Behaviour*, 2 (9), 637–44.

Carroll, Lauren. 2015. Ted Cruz's world's on fire, but not for the last 17 years, *Politifact*, March 20. https://www.politifact.com/truth-o-meter/statements/2015/mar/20/ted-cruz/ted-cruzs-worlds-fire-not-last-17-years/.

Chan, C. 2014. *Gun Deaths in Florida*, Florida Department of Law Enforcement, *Reuters*, February 16.

Chappell, Bill. 2015. Winner of French Scrabble title does not speak French, *The Two-Way: Breaking News From NPR*. https://www.npr.org/sections/thetwo-way/2015/07/21/424980378/winner-of-french-scrabble-title-does-not-speak-french.

Charpentier, Arthur. 2015. P-hacking, or cheating on a *p*-value, *Freakonometrics*. Accessed November 3, 2018. https://freakonometrics.hypotheses.org/19817.

Chatfield, Chris. 1995. Model uncertainty, data mining and statistical inference, *Journal of the Royal Statistical Society A* 158, 419–66.

Chollet, Francois. 2017. *Deep Learning With Python*. Manning Publications.

Chojniak, Rubens. 2015. Incidentalomas: Managing risks, *Radiologia Brasileira* 48 (4). https://www.ncbi.nlm.nih.gov/pmc/articles/PMC4567356/.

Christensen, Jen. 2015. What we know (and mostly don't) about Theranos' science—CNN. *CNN*. Accessed November 3, 2018, https://www.google.com/amp/s/amp.cnn.com/cnn/2015/11/12/health/theranos-what-we-know-science/index.html.

Cios, K.J. and Pedrycz, W., et al. 2007. *Data Mining: A Knowledge Discovery Approach*, New York, Springer.

Coase, R, 1988. How should economists choose?. In: *Ideas, Their Origins and Their Consequences: Lectures to Commemorate the Life and Work of G. Warren Nutter*, American Enterprise Institute for Public Policy Research.

Cohn, Carolyn, 2016, Facebook stymies Admiral's plans to use social media data to price insurance premiums, *Reuters*, November 2. Available from: https://www.reuters.com/article/us-insurance-admiral-facebook/facebook-stymies-admirals-plans-to-use-social-media-data-to-price-insurance-premiums-idUSKBN12X1WP.

Collins, Stuart. 2008. Crisis offers lessons on risk extremes, aggregation, *Business Insurance*, December 11.

Cruz, Ted Cruz. 2015. Interview, *Late Night with Seth Meyers*, March 17.

Culotta, Elizabeth and Gibbons, Ann. 2016. Almost all living people outside of Africa trace back to a single migration more than 50,000 years ago, *Science*, September 21.

Cummings, Mike. 2018. Assessing cryptocurrency with Yale economist Aleh Tsyvinski, *Yale News*, August 6.

Crawford, Kate. 2018. The hidden biases in big data. *Harvard Business Review*. January 16. Accessed November 3, 2018. https://hbr.org/2013/04/the-hidden-biases-in-big-data.

Dash, Eric. 2007. Citigroup acknowledges poor risk management, *The New York Times*, October 16. https://www.nytimes.com/2008/11/23/business/23citi.html.

Dash, Eric and Creswell, Julie. 2008. The Reckoning: Citigroup saw no red flags even as it made bolder bets, *nytimes.com*, November 23. https://www.nytimes.com/2008/11/23/business/23citi.html.

Dastin, Jeffrey. 2018. Amazon scraps secret AI recruiting tool that showed bias against women, *Reuters*, October 9. https://www.reuters.com/article/us-amazon-com-jobs-automation-insight/amazon-scraps-secret-ai-recruiting-tool-that-showed-bias-against-women-idUSKCN1MK08G.

Davis, Ernest, 2014. The technological singularity: the singularity and the state of the art in artificial intelligence, *Ubiquity*, Association for Computing Machinery. October. Available from: http://ubiquity.acm.org/article.cfm?id=2667640.

Dehghan, Mahshid and Mente, Andrew, et al, 2018, Association of dairy intake with cardiovascular disease and mortality in 21 countries from five continents (PURE): a prospective cohort study, *The Lancet*, 392 (10161), 2288–97.

Devlin, Hannah, 2015. Rise of the robots: how long do we have until they take our jobs?, *The Guardian*, February 4.

Doe, George. 2014. With genetic testing, I gave my parents the gift of divorce, *Vox*, September 9.

Duhigg, Charles, 2012. How companies learn your secrets, *The New York Times Magazine*. February 16.

Edward, Mark. 2009. Connie's conundrums, *Skepticblog*, July 13. Accessed November 3, 2018. https://www.skepticblog.org/2009/07/13/connies-conundrums/.

Eichstaedt, J.C. and Schwartz, H.A., et al. 2015. Psychological language on Twitter predicts county-level heart disease mortality, *Psychological Science*, 26 (2): 159–69. doi:10.1177/0956797614557867.

Engel, Pamela. 2014. This chart shows an alarming rise in Florida gun deaths after "Stand Your Ground" was enacted, *Business Insider*, February 18.

Esteva, Andre and Kuprel, Brett, et al. 2017. Dermatologist-level classification of skin cancer with deep neural networks, *Nature*, 542 (7639), 115–18.

Evans, R. and Jumper, J. 2018. De novo structure prediction with deep-learning based scoring. In: *Thirteenth Critical Assessment of Techniques for Protein Structure Prediction (Abstracts)* December 1–4, 2018.Accessed December 10, 2018. https://deepmind.com/blog/alphafold/.

Evtimov, I., Eykholt, K. and Fernandes, E., et al. 2017. Robust physical-world attacks on deep learning models. https://arxiv.org/abs/1707.08945.

Farmer, Brit McCandless. 2018. Could your DNA help solve a cold case?, *CBS News*. October 21. Accessed November 3, 2018. https://www.cbsnews.com/news/could-your-dna-help-solve-a-cold-case-60-minutes/.

Fayyad, U., Piatetsky-Shapiro, G. and Smyth, P. 1996. From data mining to knowledge discovery in databases, *AI Magazine*, 17 (3), 37–54.

Finlayson, Samuel G., Kohane, Isaac S. and Beam, Andrew L. 2019. Adversarial attacks against medical deep learning systems, arXiv:1804.05296v1.

Freedman, David H. 2011. Why economic models are always wrong, *Scientific American*. October 26. Accessed November 3, 2018. https://www.scientificamerican.com/article/finance-why-economic-models-are-always-wrong/.

Freedman, David H. 2013. "Survival of the wrongest", *Columbia Journalism Review*. January/February. Accessed November 3, 2018. https://archives.cjr.org/cover_story/survival_of_the_wrongest.php.

Freedman, David H. 2014. How junk food can end obesity, *The Atlantic*. February 19. Accessed November 3, 2018. https://www.theatlantic.com/magazine/archive/2013/07/how-junk-food-can-end-obesity/309396/.

Freedman, David H. 2015. Lies, damned lies, and medical science, *The Atlantic*. September 2. Accessed November 3, 2018. https://www.theatlantic.com/magazine/archive/2010/11/lies-damned-lies-and-medical-science/308269/.

Freedman, David H. 2016. The war on stupid people, *The Atlantic*. June 21. Accessed November 3, 2018. https://www.theatlantic.com/magazine/archive/2016/07/the-war-on-stupid-people/485618/.

Furrukh, Muhammad, 2013. Tobacco smoking and lung cancer, *Sultan Qaboos University Medical Journal*, 13 (3): 345–58.

Garbade, Michael J. 2018. The FIFA 2018 World Cup: machine learning predicts likely winners, *The Startup*, June 21.

Garber, Megan, 2016. When algorithms take the stand, *The Atlantic*, June 30.

Gent, Edd. 2017. AI is easy to fool—why that needs to change, *Singularity Hub*, October 10. https://singularityhub.com/2017/10/10/ai-is-easy-to-fool-why-that-needs-to-change/#sm.0000090iyjrwixdv6q3mr4u1fxlv6.

Goddard Institute for Space Studies, National Aeronautics and Space Administration. 2018. GISS Surface Temperature Analysis. https://data.giss.nasa.gov/gistemp/.

Gómez-Barris, Macarena. 2018. Decolonial Futures, *Social Text,* June 7.

Gorgolewski, Chris. 2018. Liberating Data—an Interview with John Ioannidis, *Chris Gorgolewski: Multiple Comparisons*, June 25. Accessed November 3, 2018. http://blog.chrisgorgolewski.org/2018/06/liberating-data-interview-with-john.html.

Grant, T.P. 2017. History of jiu-jitsu: coming to America and the birth of the UFC, *Bleacher Report*. October 3. Accessed November 2, 2018. https://bleacherreport.com/articles/654500-history-of-jiu-jitsu-coming-to-america-and-the-birth-of-the-ufc.

Greene, Lana. 2017. Beyond Babel: the limits of computer translations, *The Economist*, January 7. 422 (9002), 7.

Granville, Vincent. 2018. Statistical significance and p-values take another blow, *Data Science Central*, September 21. Accessed November 3, 2018. https://www.datascience-central.com/profiles/blogs/statistical-significance-and-p-values-take-another-blow.

Greenberg, Andy. 2015. The plot to free North Korea with smuggled episodes of "Friends", *Wired,* March 1.

Greenberg, Andy. 2016. Donate your old USB drives to fight North Korean brainwashing, *Wired,* February 9.

Ha, David. 1918. Reinforcement Learning for Improving Agent Design, *Google Brain*, October 10.

Hamblin, James. 2018. A credibility crisis in food science, *The Atlantic*, September.

Hands, Jack. 2016. Flashdrives for freedom? 20,000 USBs to be smuggled into North Korea, *The Guardian*, March 22.

Hartnett, Kevin and Quanta Magazine. 2018. To build truly intelligent machines, teach them cause and effect, *Quanta Magazine*, May 15. Accessed November 2, 2018. https://www.quantamagazine.org/to-build-truly-intelligent-machines-teach-them-cause-and-effect-20180515/.

Hayden, Erika Check. 2017. The rise and fall and rise again of 23andMe, *Nature*, October 11.

Hays, Constance L. 2004. What Wal-Mart knows about customers' habits, *The New York Times*, November 14. Accessed November 1, 2018. https://www.nytimes.com/2004/11/14/business/yourmoney/what-walmart-knows-about-customers-habits.html.

Herkewitz, William. 2014. Why Watson and Siri are not real AI, *Popular Mechanics*, February 10.

Hernandez, Daniela and Ted Greenwald. 2018. IBM has a Watson dilemma, *The Wall Street Journal*. August 11. Accessed November 3, 2018. https://www.wsj.com/articles/ibm-bet-billions-that-watson-could-improve-cancer-treatment-it-hasnt-worked-1533961147.

Hill, Kashmir. 2016. How Target figured out a teen girl was pregnant before her father did, *Forbes*. March 31. Accessed November 3, 2018. https://www.forbes.com/sites/kashmirhill/2012/02/16/how-target-figured-out-a-teen-girl-was-pregnant-before-her-father-did/#59d797206668.

Hope, Bradley and Chung, Juliet. 2017. The future is bumpy: high-tech hedge fund hits limits of robot stock picking, *Wall Street Journal*, December 17.

Hou, K., Xue, C. and Zhang, L. 2017. Replicating anomalies, *NBER Working Paper No. 23394*, May.

Hoerl, Arthur E. and Kennard, Robert W. 1970a. Ridge regression: biased estimation for nonorthogonal problems, *Technometrics*, 12, 55–67.

Hoerl, Arthur E and Kennard, Robert W. 1970b. Ridge regression: applications to non-orthogonal problems, *Technometrics*, 12, 69–82

Hofstadter, Douglas. 1979. *Gödel, Escher, Bach: An Eternal Golden Braid*, New York: Basic Books.

Hofstadter, Douglas and Sander, Emmanuel. 2013, *Surfaces and Essences: Analogy as the Fuel and Fire of Thinking*, New York: Basic Books.

Hvistendahl, Mara. You are a number, *Wired*, January 2018, 48–59.

Investigative. 2017. How reliable are home DNA ancestry tests? Investigation uses triplets to find out, *Inside Edition*, February 21.

Ioannidis, John P.A. 2018. All science should inform policy and regulation, *PLOS Medicine*, 15 (5). doi:10.1371/journal.pmed.1002576.

Ioannidis, John P.A. 2016. Why most clinical research is not useful, *PLOS Medicine*, 13 (6). doi: 10.1371/journal.pmed.1002049.

Ip, Greg. 2017. We survived spreadsheets, and we'll survive AI, *The Wall Street Journal*, August 3.

Jakab, Spencer. 2018. Bitcoin wasn't a bubble until it was, *The Wall Street Journal*, December 14.

James. 2018. 13 Biggest poker scandals of the last decade.", *Pokerupdate.com*. September 26. Accessed November 2, 2018. https://www.pokerupdate.com/poker-opinion/544-13-biggest-poker-scandals-last-decade.

John, Allen St. 2015. "Moneyball" makes a quiet comeback as data-driven baseball teams dominate the MLB playoffs, *Forbes*, October 19.

Jon. 2016. The world's greatest coin tosser and the survivorship bias, *Dr Wealth*. May 10, Accessed November 3, 2018. https://www.drwealth.com/the-worlds-greatest-coin-tosser-and-the-survivorship-bias/.

Jones, Ben. 2015. Avoiding data pitfalls, part 1: Gaps between data and reality, *DataRemixed*. Accessed November 2, 2018. http://dataremixed.com/2015/01/avoiding-data-pitfalls-part-1/.

Jones, Ben. 2015. Avoiding data pitfalls, part 2: Fooled by small samples, *DataRemixed*. Accessed November 2, 2018. http://dataremixed.com/2015/01/avoiding-data-pitfalls-part-2/.

Kaduk, Kevin. 2014. Broadcasters jinx Browns' turnover-free streak, *Yahoo Sports*, November 16.

Kaminski, Pete. 2011. Bodog poker moving to combat datamining, *Part Time Poker*, February 8. Accessed November 2, 2018. https://www.parttimepoker.com/bodog-poker-moving-to-combat-datamining.

Kendall, M.G. 1965. *A Course in Multivariate Statistical Analysis*. Third Edition, London: Griffin.

Kent, Christopher. 2018. AI & ophthalmology: two approaches to diagnosis, *Review of Ophthalmology*, July 11. Accessed November 2, 2018. https://www.reviewofophthalmology.com/article/ai-and-ophthalmology-two-approaches-to-diagnosis.

Keown, Alex. 2017. Meet the professor who was the first to question Theranos' research, *BioSpace*, February 21. Accessed November 3, 2018. https://www.biospace.com/article/meet-the-professor-who-was-the-first-to-question-theranos-research-/.

Kerwin, Jason. 2017. Randomization inference vs. bootstrapping for p-values, *Jason Kerwin*. September 27. Accessed November 2, 2018. https://jasonkerwin.com/nonparibus/2017/09/25/.

Khan, Roomy. 2017. Theranos' $9 billion evaporated: Stanford expert whose questions ignited the unicorn's trouble, *Forbes*. February 21.

Khomami, Nadia. 2014. 2029: The year when robots will have the power to outsmart their makers, *The Guardian*, February 22.

Knight, Will. 2016. Will AI-powered hedge funds outsmart the market?, *MIT Technology Review*, February 4.

Knight, Will. 2017. The financial world wants to open AI's black boxes, *MIT Technology Review*, April 13.

Knight, Will. 2017. There's a dark secret at the heart of artificial intelligence: no one really understands how it works, *MIT Technology Review*, April 11.

Knight, Will. 2017. Alpha Zero's alien chess shows the power, and the peculiarity, of AI, *MIT Technology Review*, December 8.

Krueger, Thomas M. and Kennedy, William F. 1990. An examination of the Super Bowl Stock Market Predictor, *The Journal of Finance*, 45 (2), 691–97.

Labi, Nadia, 2012. Misfortune Teller, *The Atlantic*, January/February 2012.

Lambrecht, Anja and Tucker, Catherine E. Algorithmic bias? An empirical study into apparent gender-based discrimination in the display of STEM career ads (March 19, 2018). Available at SSRN: https://ssrn.com/abstract=2852260 or http://dx.doi.org/10.2139/ssrn.2852260.

Lazer, David and Kennedy, Ryan, et al. 2014. The parable of Google flu: traps in big data analysis, *Science*, 343 (6176), 1203–5.

Leswing, Kif. 2016. 21 of the funniest responses you'll get from Siri, *Business Insider*, March 28.

Lee, Stephanie. 2018. Here's how Cornell scientist Brian Wansink turned shoddy data into viral studies about how we eat, *BuzzFeed*, February 25.

Lewis, Michael, 2003. *Moneyball: The Art of Winning an Unfair Game*, New York: W.W. Norton & Company.

Lewis-Kraus, Gideon. 2016. The great AI awakening, *The New York Times Magazine*, December 14.

Liptak, Adam. 2017. Sent to prison by a software program's secret algorithms, *The New York Times*, May 1.

Liu, Yukun and Tsyvinski, Aleh. 2018. Risks and returns of cryptocurrency, working paper, August 13. https://ssrn.com/abstract=3226952.

Lomas, Natasha. 2018. Apple's Tim Cook makes blistering attack on the "data industrial complex", *Tech Crunch*, October 24, 2018. https://techcrunch.com/2018/10/24/apples-tim-cook-makes-blistering-attack-on-the-data-industrial-complex/.

L.V. 2016. Spurious correlations: 15 examples, *Data Science Central*, January 26. Accessed November 2, 2018. https://www.datasciencecentral.com/profiles/blogs/spurious-correlations-15-examples.

Madrigal, Alexis. 2013. Your job, their data: the most important untold story about the future, *Atlantic*, November 21.

Maher, Amanda. 2016. Understanding Freddie & Fannie, part 1, *Buildium*, October 4. Accessed November 2, 2018. https://www.buildium.com/blog/understanding-fannie-and-freddie-part-1/.

Manzi, Jim. 2012. *Uncontrolled: The Surprising Payoff of Trial-and-Error for Business, Politics and Society*. New York: Basic Books.

Marquardt, Donald W. 1970. Generalized inverses, ridge regression, biased linear estimation, *Technometrics*, 12, 591–612.

Massy, William F. 1965. Principal components in exploratory statistical research, *Journal of the American Statistical Association*, 60, 234–56.

McDonald, Clement J. and Mazzuca, Steven A., et al. 1983. How much of the placebo "effect" is really statistical regression?, *Statistics in Medicine*, 2 (4), 417–24.

McClear, Sheila. 2019. Survey: Millennials living at home aren't saving more money, and are working less, *Ladders*, January 27, 2019. Accessed January 27, 2019. https://www.theladders.com/career-advice/survey-millennials-living-at-home-arent-saving-more-money-and-are-working-less?utm_campaign=daily-newsletter&utm_source=member&utm_medium=email&utm_content=1/29-DNL-Morningbrew&utm_term=DNL-1-29-19.

McLean, R. David and Pontiff, Jeffrey. 2016. Does academic research destroy stock return predictability?, *Journal of Finance*, 71 (1), 1540–6261.

Mole, Beth. 2018. Big nutrition research scandal sees 6 more retractions, purging popular diet tips, *ArsTechnica*, September 20.

Mollenkamp, Carrick and Ng, Serena, et al.. 2008. Behind AIG's fall, risk models failed to pass real-world test, *Wall Street Journal*, October 31.

Mutalik, Pradeep and Quanta Magazine. 2018. The slippery math of causation, *Quanta Magazine*, May 30. Accessed November 2, 2018. https://www.quantamagazine.org/the-math-of-causation-puzzle-20180530.

Nelson, Leif D. 2014. False-positives, p-hacking, statistical power, and evidential value, Berkeley Institute for Transparency in the Social Sciences, June. Accessed August 8, 2018. https://bitssblog.files.wordpress.com/2014/02/nelson-presentation.pdf.

Nelson, Robert K. and Winling, LaDale, et al. 2018. Mapping inequality, *American Panorama*, ed. Robert K. Nelson and Edward L. Ayers. Accessed November 1, 2018, https://dsl.richmond.edu/panorama/redlining/#loc=5/36.704/-96.943&opacity=0.8&text=bibliograph.

Nguyen, A., Yosinski, J. and Clune, J. 2015. Deep neural networks are easily fooled: high confidence predictions for unrecognizable images, Proceedings of the IEEE

Conference on Computer Vision and Pattern Recognition. Available from: https://arxiv.org/abs/1412.1897v4.

Nosek, Brian A. and Cohoon, Johanna, et al. 2015. Estimating the reproducibility of psychological science, *Science*, 349 (6251).

Nyst, Carly and Monaco, Nick. 2018, *State-Sponsored Trolling*, Institute for the Future. http://www.iftf.org/fileadmin/user_upload/images/DigIntel/IFTF_State_sponsored_trolling_report.pdf.

Packer, George. 2018. Ten years after the crash, *The New Yorker*, August 18.

Palast, Greg. 2018. The GOP's stealth war against voters, *Rolling Stone*, June 25, 2018.

Palmer, Donald. 2014. *Annual Report on Voter Registration List Maintenance Activities*, January 6. Accessed November 2, 2018. https://www.elections.virginia.gov/Files/maintenance-reports/2013SBEListMaintenancereport.pdf.

Parvis, Matthew. 2013. PokerNews exclusive: Isildur1 speaks about his $4 million loss to Brian Hastings, *PokerNews*, April 23. Accessed November 2, 2018. https://www.pokernews.com/news/2009/12/pokernews-exclusive-isildur1-speaks-about-the-4-million-7714.htm.

Parvis, Matthew. 2011. The $4 million controversy: Townsend admits to violations of Full Tilt Poker's terms, *PokerNews,* March 18. Accessed November 2, 2018. https://www.pokernews.com/news/2009/12/the-4-million-controversy-townsend-admits-to-violations-of-f-7711.htm.

Pearl, Judea and Mackenzie, Dana. 2018. *The Book of Why: The New Science of Cause and Effect*, New York, NY: Basic Books.

Peck, Don. 2013. They're watching you at work, *Atlantic*, December.

Pektar, Sofia. 2017. Robots will wipe out humans and take over in just a few centuries warns Royal Astronomer, *Sunday Express*, April 4.

Piatek, Clarissa. 2018. The pitfalls and potential of precision health, big data, and evidence-based medicine, *Precision Health*, April 30. Accessed November 3, 2018. https://precisionhealth.umich.edu/news-features/features/the-pitfalls-and-potential-of-precision-health-big-data-and-evidence-based-medicine/.

Piculjan, Neven. 2018. Introduction to deep learning trading in hedge funds, *Toptal*. https://www.toptal.com/deep-learning/deep-learning-trading-hedge-funds.

Popken, Ben. 2016. How $9 billion startup Theranos blew up and laid off 41%, *NBC News*, October 26. Accessed November 3, 2018. https://www.google.com/amp/s/www.nbcnews.com/news/amp/ncna671751.

Press, Gil. 2013. Big data news roundup: correlation vs. causation, *Forbes*, April 30.

Quach, Katyanna. 2017. Checkmate: DeepMind's AlphaZero AI clobbered rival chess app on non-level playing, er, board. *The Register*. December 14, 2017. Accessed December 10, 2018. https://www.theregister.co.uk/2017/12/14/deepmind_alphazero_ai_unfair/.

Reilly, Kevin. 2016. Two of the smartest people in the world on what will happen to our brains and everything else, *Business Insider*, January 18.

Reuters. 2018. Amazon ditched AI recruiting tool that favored men for technical jobs, *The Guardian*, October 10. https://www.theguardian.com/technology/2018/oct/10/amazon-hiring-ai-gender-bias-recruiting-engine.

Robbins, Bruce and Ross, Andrew. 2000. Response: mystery science theatre. In: *The Sokal Hoax: The Sham that Shook the Academy*, ed. Alan D. Sokal, University of Nebraska Press, pp. 54–8.

Rodriguez, Julio. 2009. Online poker–Full Tilt Poker suspends Brian Townsend, *Card Player*, December 22. Accessed November 2, 2018. https://www.cardplayer.com/poker-news/8201-online-poker-full-tilt-poker-suspends-brian-townsend.

Rodriguez, Julio. 2009. Online poker—the data mining dilemma, *Card Player*, December 28. Accessed November 2, 2018. https://www.cardplayer.com/poker-news/8221-online-poker-the-data-mining-dilemma.

Ross, Julie A. 1998. Brain tumors and artificial sweeteners? A lesson on not getting soured on epidemiology, *Medical and Pediatric Oncology*, 30 (1).

Ruddick, Graham. 2016. Admiral to price car insurance based on Facebook posts, *The Guardian*, November 1.

Rudgard, Olivia. 2016. Admiral to use Facebook profile to determine insurance premium, *The Telegraph*, November 2.

Rutkin, Aviva Hope. 2017. The tiny changes that can cause AI to fail, *BBC Future*, April 17.

Ryan, Rich. 2011. Data mining: to mine or not to mine, *PokerNews*, April 16. Accessed November 2, 2018. https://www.pokernews.com/news/2011/04/data-mining-online-poker-10211.htm.

Sagiroglu, S. and Sinanc, D. 2013. Big data: a review, *Proc. IEEE Int. Conf. Collaboration Technol. Syst. (CTS)*, May 2013.

Salmon, Felix and Stokes, Jon. 2010. Algorithms take control of wall street, *Wired*, December 27.

Sample, Ian. 2018. Google's DeepMind predicts 3D shapes of proteins. *The Guardian*. December 2, 2018. Accessed December 10, 2018. https://www.theguardian.com/science/2018/dec/02/google-deepminds-ai-program-alphafold-predicts-3d-shapes-of-proteins.

Seife, Charles. 2013. 23andMe is terrifying, but not for the reasons the FDA thinks, *Scientific American*, November 27.

Sharif, Mahmood and Bhagavatula, Sruti, et al. 2016. Accessorize to a crime: real and stealthy attacks on state-of-the-art face recognition, Proceedings of the 2016 ACM SIGSAC Conference on Computer and Communications Security, 1528–40.

Silver, D. and Hubert, T., et al. 2018. A general reinforcement learning algorithm that masters chess, shogi, and Go through self-play. *Science*362, no. 6419: 1140–4. doi: 10.1126/science.aar6404.

Simonite, Tom. 2017. How to upgrade judges with machine learning, *MIT Technology Review*, March 6.

Simonite, Tom. 2018. AI has a hallucination problem that's proving tough to fix, *Wired*, March 9.

Šimundić, Ana-Maria. 2013. Bias in research, *Biochemia Medica*, 23 (1), 12–15.

Singh, Angad. 2015. The French scrabble champion who speaks no French, CNN, July 22.

Smith, Gary. 1980. An example of ridge regression difficulties, *The Canadian Journal of Statistics*, 8 (2) 217–25.

Smith, Gary. 2014. *Standard Deviations: Flawed Assumptions, Tortured Data, and Other Ways to Lie with Statistics*, New York: Overlook Press.

Smith, Gary. 2017. *What the Luck?* New York: Overlook Press.

Smith, Gary. 2018. *The AI Delusion*. Oxford: Oxford University Press.

Smith, Gary and Campbell, Frank, 1980. A critique of some ridge regression methods, *Journal of the American Statistical Association*, with discussion and rejoinder, 75 (369), 74–81.

Smith, Rich. 2018. The Loser, *The Stranger*, May 23. https://www.thestranger.com/features/2018/05/23/26381494/the-loser.

Sokal, A. 1996. Transgressing the boundaries: Towards a transformative hermeneutics of quantum gravity, *Social Text*, 46/47, 217–52.

Somers, James. 2013. The man who would teach machines to think, *The Atlantic*, November.

Staff writer. 2008. Banks, not insurers, need tighter U.S. laws says Hartwig, *National Underwriter*, November 21.

Stanley, Jason and Vesla Waever, 2014. Is the United States a racial democracy?, *The New York Times*, January 12.

StarTalk. *National Geographic*. November 19, 2018.

Steering Committee of the Physicians' Health Study Research Group. 1988. Preliminary report: findings from the aspirin component of the ongoing Physicians' Health Study, *New England Journal of Medicine*, January 28, 262–64.

Strong, Steven and Whiteley, Andy. 2013. DNA evidence debunks the "out-of-Africa" theory of human evolution, *Wake Up World*, December 15.

Su, Jiawei, Vargas and Danilo Vasconcellos, et al. 2017. One pixel attack for fooling deep neural networks, Available from: https://arxiv.org/abs/1710.08864v2.

Sumagaysay, Levi. 2018. Time's up for Theranos: blood-testing company to dissolve, *The Mercury News*. September 5.

Superchi, Cecilia and Blanco, David. 2017. Interview with Professor John Ioannidis, *Methods in Research on Research*, December 11. Accessed November 3, 2018. http://miror-ejd.eu/2017/11/13/interview-with-professor-john-ioannidis/.

Symon, Evan V. 2017. Inside the shady world of DNA testing companies, cracked, December 4.

Szegedy, Christian and Zaremba, Wojciech, et al. 2014. Intriguing properties of neural networks, Google, February 19. Available from: https://www.researchgate.net/publication/259440613_Intriguing_properties_of_neural_networks

Tabarrok, Alex. 2015. The small schools myth, *Marginal Revolution*, June 10. Accessed November 3, 2018. https://marginalrevolution.com/marginalrevolution/2010/09/the-small-schools-myth.html.

Tal, Aner and Wansink, Brian. 2014. Blinded by science: trivial scientific information can increase our sense of trust in products, *Public Understanding of Science*. 25, 117–25.

Tashea, Jason. 2017. Courts are using AI to sentence criminals. That must stop now. *Wired*, April 17.

Thomas, Jace. 2007. A basic guide to data mining in online poker, *Part Time Poker*, November 6. Accessed November 2, 2018. https://www.parttimepoker.com/a-basic-guide-to-data-mining-in-online-poker.

Topol, Eric J. and Ioannidis, John P.A. 2018. Ioannidis: most research is flawed; let's fix it, *Medscape*. June 25. Accessed November 3, 2018. https://www.medscape.com/viewarticle/898405#vp_4.

Tullock G. 2001. A comment on Daniel Klein's "A plea to economists who favor liberty," *Eastern Economic Journal*, 27 (2), 203–7.

Vincent, James. 2018. DeepMind's AI can detect over 50 eye diseases as accurately as a doctor, *The Verge*, August 13. Accessed November 3, 2018. https://www.theverge.com/2018/8/13/17670156/deepmind-ai-eye-disease-doctor-moorfields.

Vorhees, William. 2016. Has AI gone too far? Automated inference of criminality using face images, *Data Science Central*, November 29.

Wainer, Howard and Zwerling, Harris L. 2006. Evidence that smaller schools do not improve student achievement, *The Phi Delta Kappan*, 88 (4), 300–3.

Wainer, Howard. 2011. *Picturing the Uncertain World: How to Understand, Communicate, and Control Uncertainty through Graphical Display.* Princeton, NJ: Princeton University Press.

Wang, Min Qi and Yan, Alice F., et al. 2018. Researcher requests for inappropriate analysis and reporting: A U.S. survey of consulting biostatisticians, *Annals of Internal Medicine*, 169 (8): 554–8.

Weaver, John Frank. 2017. Artificial intelligence owes you an explanation, *Slate,* May 8.

Weihrauch, M.R. and Diehl, V. 2004. Artificial sweeteners—do they bear a carcinogenic risk?, *Annals of Oncology*, 15 (10), 1460–5.

Wiecki, Thomas and Campbell, Andrew, et al. 2016. All that glitters is not gold: comparing backtest and out-of-sample performance on a large cohort of trading algorithms, *Journal of Investing*, 25 (3), 69–80.

Willer, Robb. 2004. The Intelligibility of Unintelligible Texts. Master's thesis. Cornell University, Department of Sociology.

Willsher, Kim. 2015. The French scrabble champion who doesn't speak French, *The Guardian*, July 21.

Winograd, Terry. 1972. Understanding natural language, *Cognitive Psychology*, 3, 1–191.

Wolchover, Natalie and Quanta Magazine. 2017. New theory cracks open the black box of deep learning. *Quanta Magazine.* Accessed December 10, 2018. https://www.quantamagazine.org/new-theory-cracks-open-the-black-box-of-deep-learning-20170921/.

Wolfson, Sam. 2016. What was really going on with this insurance company basing premiums on your Facebook posts?, *Vice*, November 2.

Woodward, James, 2016. Causation and manipulability, *The Stanford Encyclopedia of Philosophy*, Winter 2016.

Woollaston, Jennifer. 2016. Admiral's firstcarquote may breach Facebook policy by using profile data for quotes, *Wired UK*, November 2.

Wu, Xiaolin and Zhang, Xi. 2016. Automated inference on criminality using face images, Shanghai Jiao Tong University, November 21. Available at: https://arxiv.org/abs/1611.04135v1.

Wu, Xiaolin and Zhang, Xi. 2017. Responses to critiques on machine learning of criminality perceptions, Shanghai Jiao Tong University, May 26. Available at: https://arxiv.org/abs/1611.04135v3.

Yuan, Li. 2017. Want a loan in China? Keep your phone charged, *The Wall Street Journal*, April 6.

Yudkowsky, Eliezer. 2008. Artificial intelligence as a positive and negative factor in global risk. In *Global Catastrophic Risks*, eds. Nick Bostrom and Milan M. Ćirković, New York: Oxford University Press, 308–45.

Yong, Ed. 2018, A popular algorithm is no better at predicting crimes than random people, *The Atlantic*, January 17.

Zeki, Semir and Romaya, John Paul, et al. 2014. The experience of mathematical beauty and its neural correlates, *Frontiers in Human Neuroscience*, February 13. 8, article 68.

Zuckerman, Gregory and Hope, Bradley. 2017. The quants run Wall Street now, *The Wall Street Journal*, May 21.

INDEX

Abramoff, Michael 160–3
ABTester 155–6, 158
AIG 234–5, 238
Alderson, Sandy 29–30
algorithmic criminology
    197–200
AlphaGo 98, 203, 205
AlphaGo Zero 203
AlphaZero 203, 205
Amazon 207–8
Anderson, Chris 35, 158–9
Andrabi, Tahir 150
anthropomorphization 86–7
Apple 223
Arnold, John 119
aspartame 122–3
aspirin and heart attacks 34
Attenborough, David 27
Australia 220

Babbage, Charles 85
backgammon 19, 201–2
baseball regression 186–8
bean machine 184
Beane, Billy 27–31
Bear Sterns 238
beer and marriage 156–7
Bem, Daryl 114–15, 117
Bernstein, Peter 176
BestWeb 173–4, 177–8, 184–6
Bhambra, Gagan 132–4
Big Brother 226
bingo 92–3
binomial distribution 78–82
bitcoin bubble 57–63
black box 87–8, 99, 101, 162–3,
    166, 197, 199, 208–9
Black, Fischer 69
Black–Scholes model 69

black swan 68
Blom, Viktor 163–4
Boaty McBoatface 26–7
Bock, Laszlo 206–7
bottomless bowls 111–12
bowling streaks 214–6
Brazilian Jiu-Jitsu 145–9
break-even theory 166–8
British Open 191–2
British Royal Air Force 24–5
broadcaster jinx 192–3
Brown, Derren 115–16, 148
Brown, Nick 113, 117–18
Buffett, Warren 66, 193, 233
BuyNow 121–2

Calude, Cristian S. 157
Cameron, William Bruce 207
cancer 19, 37–41, 118,
    122–3, 129
cane toads 220
CAPTCHAs 103–4
Casey, Kathleen 236–7
cell phones and brain
    cancer 37–41
Center for Open Science 120
checkers 202–3
cherry picking 35
chess 203–6
Chicago crime data 5–6
Chinook 202
cigarettes and lung cancer 38–9
Citigroup 234, 239
climate change 21–2
Coase, Ronald 115
coffee and health 118–19
coin streaks 91–2
collateralized debt obligations
    (CDOs) 232–33

COMPAS 197–200
computer-generated
    code 93–8
computer poker 88–9
Cook, Tim 223–4
correlation versus
    causation 158–9
Coyne, James 117–18
credit default swap
    (CDS) 232–3
crime by day 5–6
criminal faces 210
Cruz, Ted 21–2

Das, Jishnu 150
data clown 2, 153
data dredging 35
data exploration 35
data–industrial complex 223
data mining 34–6, 61
data-mining heart attack
    data 47–9
data scientist 1–2
data snooping 35
Davie, Art 145, 148–9
Davis, Ernest 102
Decline effect 190
DeepMind 203
deep neural networks
    (DNNs) 98–9
derivatives 232–4
diabetic retinopathy 160–3
diversification 230–2
DNA testing 216–19
DomainsAreUs 220–2
Dorsey-Palmateer, Reid
    214–16
Dr. Fox Effect 65–6
Dwan, Tom 163–4

earthquakes 4
Economist 2, 35
Eisenhower, Dwight D. 223
England loves teal 51–2
Equivalent 187
ESP hacking 123–5
Etzioni, Oren 103
Evolving Artificial Intelligence
    Laboratory 106
extrasensory perception
    (ESP) 123–4

Facebook and burglaries 43–5
false positives 116
Fannie Mae 229
fat tails 67–9, 235
Federal Reserve 7, 237–9
Ferguson, Chris 139, 141
Feynman, Richard 37, 130
Feynman Trap 37, 41, 63
file drawer effect 120, 124
Firth, J.R. 102
fishing expeditions 35,
    114–15, 117
FiveThirtyEight 20
Florida stand-your-ground
    law 213–14
flu outbreaks 41–2
Fox, Michael 65
Freddie Mac 229
freedom of expression
    224–6
Frum, David 11–13
Fuld, Richard 239
Full Tilt Poker 166–8

Galton, Francis 183–4
gambler's fallacy 175
Game of Thrones 102
garbage in, garbage out 67
genetic testing 216–19
Gilovich, Thomas 214
Ginnie Mae 229
global warming 21–2
Go 203
gold 19–21
Golden State Killer 219
Goldman Sachs 239
golf 13–16

Google 1, 45, 60, 101, 104, 107,
    158–9, 206, 225
Google Assistant 199–200
Google Flu 41–2, 45
Google Translate 100–1
Gracie, Rorion 145–9
graphs, misleading 211–16
Great Chicago Fire 234
Great Depression 208–9
Great Recession 229, 237–9
Greely, Hank 219

Hassabis, Demis 203
Hastings, Brian 163–4
Henry, John 30
hi-tech redlining 208–10
hindsight bias 144
Hofstadter, Douglas 105
Home Owners' Loan
    Corporation (HOLC) 208–9
hospital readmissions 8–10
hot streaks 91, 214–16
household wealth error 7
hurricanes 114, 170–1

Ig Nobel award 112
image recognition 104–8
incidentalomas 128
in-sample data 46–7
information harvesting 35
Interphone study 40–1
Ioannidis, John 127–31
Ip, Greg 72
Isildur1 163
Israeli flight instructors
    189–90

J.P. Morgan 238–9
James, Bill 28–30
James Randi Educational
    Foundation 125
Jellyfish 202
Jeopardy! 201
job applications 206–8
Jong-un, Kim 225–6

Kahneman, Daniel 13, 166,
    189–90
Kelly criterion 139–40

Kelley, Truman 181
Kelley's equation 181–2
Keynes, John Maynard 29, 76
Kharitonov, Michael 50–1
Khwaja, Asim Ijaz 150
King Gustav III 118–19
kitchen-sink approach 70–3
knowledge extraction 35
Krabbé, Tim 204
Kristof, Nicholas 150

Lafferty, Don 202
language translation 99–101
Laura and John Arnold
    Foundation 119
law of averages 175
law of small numbers 13–15
Layne, Bobby 89
leap year 3–4
LEAPS 150–2
Lehman Brothers 238
Lewis, Michael 30
Long Island Iced Tea 58
Longo, Giuseppe 157
Loomis, Eric 197–200
luck in golf 13–16

Madden Curse 192
Maisie 101
Major League Baseball 27–30
marital bargaining power
    168–9
Masters golf tournament 14–16
Matrix 86
McAuliffe, Jon 50–1
Merrill Lynch 238
Merritt, Deborah 65
Meta-Research Innovation
    Center 128
Meyers, Seth 21
Mickelson, Phil 14
military–industrial
    complex 223
military veterans 6–7
Millennials 156
Moneyball 30
Monte Carlo simulations 89–91
mortgage pools 229–32
musician mortality 9–11

Musk, Elon 86–7
mutual funds 23–5

National Basketball
  Association (NBA) 17
National Security Agency
  (NSA) 225
NINJA loans 230–1
nOCD 131–5
nonlinear models 73–5
North Korea 225–6
Northpointe 197
Nosek, Brian, 119–20
nutritional studies 118–19

Oakland Athletics 29–30
Octagon 144–9
Office 93
OKCupid 210
One Million Dollar
  Paranormal Challenge 125
Open Practice Badges 120–1
optical character recognition
  (OCR) 104
out-of-sample data 46–7
overfitting data 70–3

p-hacking 35, 61, 113, 115–17,
  120, 122–4, 127, 152, 186, 236
p-value 35–6, 55, 78–9, 81,
  115–16, 131
Pakistan 150–2
Pearl, Judea 159
pizza papers 113–14
Pizza Principle 123
placebo effect 190–1
Pomona College 132–5
poker 88–9, 135–43, 163–4,
  166–7
poker, the System 135–41
pop-tarts 170–1
pregnancy predictor 42–3
presidential approval
  ratings 194
principal components
  regression 82–3
propranolol 182–3
Prudential Insurance 22–3
publication effect 120, 124

QuickStop 51–5

radiation 37–8
Randi Prize 125
Redleaf, Andrew 239
redlining 208–9
replication crisis 117,
  119–22, 130
Reproducibility Project 119
Regression fallacy 191
regression toward the
  mean 120, 174–9
Retinator 160–3
retirement planning 89–91
Rhine, J.B. 124–5
Richards, Nigel 100
right to privacy 223–4
Robo-Tester 121–2
Rogan, Joe 148
Ross, Julie A. 123
Rossi, Dino 176

sabermetrics 27–31
Samuelson, Paul 230–1
Schaeffer, Jonathan 202–3
Schmidt-Erfurth, Ursula 162
Scholes, Myron 69
school bus image 105–7
school performance 19
Scholastic Aptitude Test
  (SAT) 180–3
scientific method 34–5, 129,
  159, 170
self-selection bias 26–7, 151
sepsis patients 8–10
set-aside solution 45–7
shortfall risk 90–1
short-stack strategy 136–41
Silver, Nate 20
Simpson's paradox 52–5
Siri 200
skeletal essence 105
small samples 17–19
Smith, Steven 131–5
smoking and cancer 129
Snowden, Edward 225
Snowie 202
Socratic Paradox 128
soul mates 194–5

South, Cole 163–4
spiders and Spelling Bee
  words 40
Spies, Jeffrey 119
Sports Illustrated 37
Sports Illustrated jinx 192
Standard & Poor's 229, 236
statistical overconfidence 78–82
stepwise regression 55–6
stock market 1987 crash 7, 68
stock market fat tails 67–9
stock market overfitting
  data 75–8
stock market and Twitter 50
stop signs 106–8
Streisand, Barbra 222
Streisand effect 222
Super Bowl indicator 69
Super Bowl predictions 69–72
survivorship bias 24–5, 115
Swift, Jane 206–7

tanks and clouds 160
Target 42–3
TD-Gammon 202
Terminator 86
Texas Sharpshooter
  Fallacy 36–7, 49, 81, 125
Them That Got 206
theory before data 63
Theranos 127–9
ThinkNot 50
Thompson, Klay 193–4
Tinsley, Marion 202
Townsend, Brian 163–4
too big to fail 237
TradeALot 75–8
training data 46
tranches 232
TryAnything 33, 52
TrySomething 152–3
tulip bulb bubble 58
Tversky, Amos 13, 166, 214
tweets and heart attacks 117–18
Twitter 1, 45, 50–1, 60, 117–18

Ultimate Fighting
  Championship 144–9
underwater mortgages 231–2

unintended consequences 220
up is up 33, 34, 43, 45, 48, 50,
    42, 153, 156, 168, 170

validation data 46
Vallone, Robert 214
Voleon Group 50–1
voter name crosschecks
    210

Wainer, Howard 179–80
Wal-Mart 170

Wald, Abraham 24–5
Wansink, Brian 111–14
Watson 201
wealth inequality 11–13
Wells Fargo 239
WhatWorks 121–2
wheat pennies 143–4
Williams, Ted 27
Winograd, Terry 102
Winograd Schemas 102–3
wins above replacement
    (WAR) 31

Wired 35, 158
Wolfram Image Identification
    Project 106
World Cup predictions
    75
World War II planes
    24–5

Yap 57
Young, Cy 27

Zener cards 124